WHERE TO GO FOR WILDLIFE

in Berkshire, Buckinghamshire & Oxfordshire

Edited by Geoffrey Young

BBONT

Berkshire,
Buckinghamshire &
Oxfordshire
Naturalists'
Trust

3 Church Cowley Road, Rose Hill, Oxford OX4 3JR Tel: Oxford 775476

First published by BBONT 1989

British Library Cataloguing in Publication Data
Berkshire, Buckinghamshire and Oxfordshire Naturalists' Trust
Where to go for wildlife.
1. England. Nature reserves
I. Title II. Young, Geoffrey, *1936–*
639.9'0942

ISBN 0-9501625-6-6

Designed by Peter Creed, Pisces Publications, Oxford
Typeset by DMD, Oxford
Printed by Bocardo Press, Didcot

Cover photograph: Orange-tip on primrose, BBONT

CONTENTS

ACKNOWLEDGEMENTS

I would like to thank the following people for their help with the production of this book:

Michael Horwood, who wrote the BBONT Nature Reserves Handbook on which many of the reserve descriptions are based; Peter Creed for supervising the layout and production of the book; Dr. Peggy Varley, Dr. David Steel, Fiona Danks, Stella Waterman and Peter Creed for making many helpful suggestions for editing the text; Joe Little for drawing the maps; Peter Creed, Cherry Fairley, Edith Golnast, Charlotte Matthews, Denys Ovenden and Sue Sandys for the drawings.

Geoffrey Young
February 1989

FOREWORD

"To promote wildlife conservation in Berkshire, Buckinghamshire and Oxfordshire by acquiring and managing nature reserves" is one of the Trust's main objectives. Since it was set up in 1959, BBONT has acquired almost 100 reserves in three counties, with a total area of about 1150 hectares (2800 acres). Some have been bought, with considerable assistance, for which we are very grateful, from national and international organizations, local authorities, charitable Trusts and many private individuals. Others are held on lease or licence and BBONT manages them for the conservation of wildlife.

BBONT owns and looks after reserves for the benefit of ourselves and our children. The pressures of increasing population, pollution and advanced agricultural technology have led to the disappearance of beautiful and interesting plants and animals from the British countryside because of destruction of their habitats. By conserving habitats, we protect many species and provide a mosaic of diverse environments scattered through our "tamed" countryside. We want people to visit these habitats and enjoy sights and sounds which are becoming increasingly rare.

This book is written for the general membership of the Trust – for people who are not specialists in the sciences of ecology, botany, entomology or ornithology but who are interested in wildlife and natural environments. If you have specialist knowledge and interests, you can obtain more detailed information from BBONT Offices. For each of the reserves, the author explains what sort of habitats are there and describes the special features of that site. This sometimes leads him to write about a general topic, relevant to several reserves – the life histories of certain butterflies; possible explanations for why certain plants and animals have very restricted distributions; medieval methods of agriculture and forestry; and so on. So dipping into the book can provide odd "snippets" of information about countryside matters.

We hope that members will use the book and its maps to plan visits to the reserves so that they go at an appropriate time of year and are ready to look out for the interesting features which are the reason why we acquired and wish to conserve that particular place. We also hope that people who are not BBONT members will buy the book and visit some of our reserves – and that their enjoyment will stimulate them into joining the Trust.

The reserves are scattered through the three counties and include a great diversity of habitats and plant and animal communities. Some are very fragile environments and therefore access to them is restricted; if you wish to visit one of these, you must obtain a permit from BBONT HQ in Oxford. The majority are open to the public but they are all special places and we ask visitors to avoid damage and behave sensibly. Noisy groups would frighten birds, deer and other wildlife and spoil places for other visitors. In some reserves, you should keep to the paths, either through the whole year or at times when birds may be nesting. Dogs must be kept on leads on many reserves and especially when sheep or cattle are grazing. These animals are "management tools"; their feeding promotes the survival of colourful wildflowers and their associated butterflies. You may see other signs of management such as newly-coppiced woodlands, laid hedges or ponds cleared of debris.

Conservation is an active process. We have to manage our reserves to keep them in the particular state of natural succession that is of special interest today. We welcome active volunteers to help on BBONT reserves and we are very grateful for financial contributions to support our work in conservation and in environmental education for the community. Enjoy your visits and please join us in protecting and conserving wildlife and beautiful, interesting habitats in our three counties.

Peggy Varley, Chairman of BBONT Council

WHAT IS BBONT?

The Berkshire, Buckinghamshire and Oxfordshire Naturalists' Trust is one of a countrywide network of County Wildlife Trusts aiming to promote wildlife conservation. It is a registered charity and one of its most important activities is to acquire and manage nature reserves, which are areas of high wildlife value.

The pressures that are put on our countryside by modern farming, forestry and development severely threaten places like this; once destroyed such sites cannot be re-created. When ancient meadows or woodlands are devastated they have gone for ever; conserving such sites is not only important for the wild plants and animals living there but is also essential for the long term well being of our own species. The natural environment is a resource that we cannot afford to destroy; it provides us with fuel, medicines, food and many other valuable products. The complexity of natural habitats is such that nobody fully understands how they work but we can often see when the natural balance is upset; plant and animal indicator species can warn us of the effects of certain substances or activities, eg; aquatic invertebrate communities can be used to measure water pollution.

It is because so many wildlife habitats have been and are continuing to be destroyed that there is a need for organisations such as BBONT. Like the other County Trusts we are involved in protecting local wildlife habitats and in encouraging people to learn about and enjoy our unique wildlife heritage.

AN INTRODUCTION TO NATURE RESERVES

BBONT's 90+ nature reserves are distributed widely over the three counties; these reserves are key sites with high wildlife value and it is essential to manage them so that the quality of their habitats is maintained and enhanced. We aim to conserve a wide variety of habitats including mixed deciduous woods, chalk downland, wet meadows, hay meadows, bogs and heaths, hedgerows and even disused railway lines and gravel pits. Buying and/or leasing and managing nature reserves costs money; we often receive grant aid to buy specific sites but you can help by joining BBONT because we rely largely on membership subscriptions and fundraising for our income.

Why do we need nature reserves?

The British countryside has evolved alongside human populations for thousands of years. In previous centuries there was relative harmony between people and their environment and changes took place slowly. Plant and animal communities were modified gradually by the various systems of management. Since the 19th century very rapid changes have taken place; towns and cities have grown as the population has increased, industry has developed and transport networks have become more complex.

But perhaps the most dramatic changes to our countryside have occurred because of developments in agriculture and forestry; the small fields, hedgerows and coppiced woodlands of yesterday have in many cases been replaced by large fields with monocultures of arable crops, sprayed with chemicals and cultivated by huge machines. These developments have occurred because of pressures on the farming community to increase food production. Change has been rapid and we have certainly benefitted from the increases but the diversity of plants and animals in the countryside has been seriously depleted in the process. In an effort to look after what remains, we have had to create nature reserves but if conservation is to be effective we must address ourselves to the lack of ecological understanding in people's everyday lives and provide opportunities for them to enjoy and understand the natural world. BBONT hopes that this publication will ensure that the reserves it holds in trust can contribute towards improving that understanding.

The wildlife areas that we are are now trying to conserve all depend on human interference for their survival; ever since the primaeval forest was first cut down we have managed the land in some way or other. Different farming techniques which took account of soil type and drainage have led to a wide variety of plant and associated animal communities that modern monoculture can rapidly eradicate. This means that practically none of today's landscape is entirely "natural" – it has all been influenced by people, so if we wish to continue to enjoy a rich and varied countryside we must manage our wildlife habitats sensitively and with reference to the traditional management methods which brought them about.

The countryside then . . .

. . . and now.

Looking after nature reserves

We need to manage nature reserves because, left to themselves, they would change as a result of natural processes. If a field of grass is not cut or grazed, tall vigorous grasses overshadow and smother the smaller plants including many wildflowers. Seeds from neighbouring woods and hedgerows will be dispersed by wind or animals and some of those landing in the grassland will germinate and grow; the area will firstly become scrub and eventually woodland. This change from grassland to woodland is an example of the natural process called SUCCESSION; the woodland is the CLIMAX vegetation. The field of grass was probably created many years ago by clearing ancient woodland to provide food for livestock and it represents a SUB-CLIMAX vegetation held in check by a constant management regime.

The aim of conservation management is to maintain a wide variety of appropriate habitats and species at each site and to ensure that the natural habitats of these three counties continue to survive. The difference between how we manage these sites now and how they were managed in the past is that we are now managing them in order to conserve the wild species that live there as a result of historic practices in forestry and agriculture.

Unmanaged grassland changes into scrub then into woodland.

How do we manage our reserves?

1. **Hay meadows** eg; Bernwood Meadows (Bucks)
   Unimproved hay meadows support a wide variety of plant and associated insect species; when such meadows are "improved" by ploughing, draining, fertilising or re-seeding, their interest is lost because the resulting increase in fertility encourages the tougher species to dominate over the less aggressive wildflowers and grasses.

   We manage hay meadows by cutting for hay between July and September; the hay is removed to reduce fertility and to prevent an accumulating blanket of rotting vegetation. Livestock are often put in to graze the aftermath; as selective grazers they will help to maintain plant diversity and their footprints provide exposed earth for annual plants to germinate readily.
2. **Chalk grassland** eg; Hartslock (Oxon)
   Chalk grasslands occur on thin, dry calcareous soils and often include a great range of wildflowers and insects; such grasslands have been traditionally managed by the grazing of sheep or rabbits. BBONT has a flock of more than 150 sheep that are moved around our various grassland reserves, acting as our living lawnmowers. They do a better job than lawnmowers; sheep graze selectively but lawnmowers are indiscriminate.

3. **Deciduous woodlands** eg; Moor Copse (Berks)
A very great variety of woodland reserves has arisen through various systems of management. The most valuable woods in terms of wildlife are ancient woods that have existed for several centuries and possibly since the last Ice Age; these are where one finds carpets of many different wildflowers in the Spring. Most ancient woods were managed as "coppice with standards" which produced a wide range of sizes of timber suitable for various uses.

Coppicing benefits wildlife in various ways; when an area of woodland is first coppiced much more light reaches the woodland floor encouraging growth of the ground flora such as bluebells, primroses and yellow archangel. As shoots grow from the coppiced stool, they provide food for caterpillars and other insects and eventually there are nesting sites for birds, including the nightingale. When the coppice has grown tall and dense it will restrict the light reaching the woodland floor. Traditionally a wood would have been divided into compartments with a different section being coppiced each year on a seven to fifteen year cycle, thus ensuring a constant supply of coppice products. An actively coppiced wood is beneficial to wildlife because all stages of the succession from open ground to thicket and old coppice are represented, thus providing habitats for a great variety of plant and animal species.

Traditional products of a coppiced woodland.

Woodland management tasks include:

a. Coppicing.
b. Clearance and maintenance of footpaths.
c. Removal of introduced species eg: snowberry, sycamore. Such non-native species are often of low wildlife value and tend to spread rapidly if not controlled. Sycamore is a particularly invasive species in woodlands where it casts very dense shade so that few ground plants are able to grow.

   Many true native plant species are of high wildlife value because they tend to support more invertebrates; those tree species that have only been here for a few hundred years do not usually have so many invertebrates associated with them because there has not been enough time for them to adapt or evolve.
d. Creation of rides to allow more light to reach the ground; the plants that are thus encouraged attract butterflies and other animals. Rides are managed as grasslands to provide diversity.

   A woodland ride is similar to a woodland edge, representing a border between two habitats; such areas are known to hold a rich variety of wildlife.
e. Maintenance of an uneven-aged stand of trees to ensure that young trees are present to replace the older trees as they die.
f. Harvesting coniferous trees which have been planted on some of the sites that are now BBONT reserves. Removing these trees can improve the wildlife value of a site, especially if it is replanted with indigenous species, and also provides revenue for the Trust.
g. Sometimes there is a policy of non-intervention where a wood is allowed to develop naturally to become a future "wildwood".

4. **Ponds** eg; Henry Stephen/C. S. Lewis Nature Reserve (Oxon)
Most ponds have been created to provide water for livestock or through extraction of stone, gravel or clay. A pond is a temporary landscape feature and naturally becomes filled with dead plant matter and silt; left for long enough it will become a marsh and eventually dry out completely. Ponds provide valuable habitat for amphibians and many invertebrates and birds. Managing ponds entails occasional dredging, limiting plant growth and cutting back of overhanging trees.

5. **Heathland** eg; Inkpen Common (Berks)
Heathland is the type of vegetation usually found on poor, acidic soils; it is dominated by heathers (eg: Ling, bell heather) with a variety of fine grasses and wildflowers. Natural succession leads to invasion by birch followed by other tree species.

Traditionally many heathland areas were managed as common land and local people had common rights to graze livestock, cut wood and peat for fuel and gather bracken and heather for bedding. Another traditional method of management was burning; when done carefully this has the effect of rejuvenating the heather. Management of these areas today entails controlling birch and also removing bracken which tends to spread and dominate heathland. Some areas of scrub are left alone in order to provide cover for birds.

6. **Bogs** eg; Owlsmoor Bog and Heath (Berks)
   Bogs are wet areas that develop on poorly drained soils and are dominated by sphagnum mosses; some interesting insectivorous plants may grow among them. There are now very few bogs in our three counties. Management entails removal of invading scrub (often birch) and sometimes digging out damp areas to prevent them from drying out. Low nutrient levels are essential and the ecology of bogs can be greatly altered by the presence of fertilizers.
7. **Hedgerows** eg; Tuckmill Meadow (Oxon)
   As more areas of woodland have disappeared from the countryside, hedgerows have become increasingly important, providing valuable habitats for nesting birds, small mammals and insects. Hedgerows are linear "woodland edge" habitats acting as corridors between patches of woodland and as refuges for some species that may forage in open fields, such as predatory insects that eat aphids. Traditional management entailed laying hedges about once every fifteen years; this activity promotes new growth and in the long term provides a thick, stockproof hedge that is also very valuable to wildlife. Modern cutting techniques tend to prevent the development of fruits and also of climbing plants which can become established at the end of the cutting cycle. Coppicing is another form of traditional hedge management.

A neglected hedge.

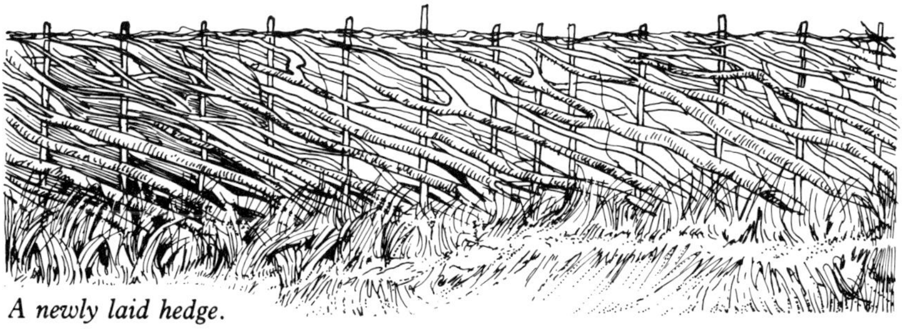

A newly laid hedge.

8. **Man-made wildlife sites** eg; Weston Turville Reservoir (Bucks)
Not all our nature reserves have arisen through traditional management over many years; some have been created through much more recent and seemingly destructive human activities. Our list of nature reserves includes ponds created through clay extraction for brickmaking, old railway lines, an old sand and gravel quarry and disused gravel pits. When such sites are left undisturbed for a number of years they can develop into valuable habitats of great wildlife interest. It is possible to create particular types of habitat by appropriate engineering followed by planting and seeding. Such sites are not only of scientific interest, providing good opportunities to study the early stages of succession, but also provide local amenities.

Who manages the BBONT nature reserves?

Managing nature reserves is a considerable responsibility and involves a great deal of time, expertise and money. Most of the work is done by volunteers; each reserve has a volunteer Reserve Manager and a team of Wardens who keep an eye on the site and carry out any necessary management work. These volunteers have a vital role to play in the running of BBONT and the maintenance of its estate. Management plans and schedules of work over the year are prepared by experts with the help of local members.

A small staff is employed by BBONT, some of whom are involved in reserve management. Three County Officers are employed, each one being responsible for the reserves in their particular county. There are two Field Officers, one responsible for the care of our flock of sheep and the grassland sites and the other to oversee the management of our woodlands.

What else does BBONT do?

BBONT's major responsibility is viewed by many as the establishment and management of nature reserves, ensuring the long term survival of some of the most valuable wildlife sites in the three counties. However, the Trust also has a responsibility for the wider countryside and for increasing public awareness of the threats to our wildlife heritage; nature reserves represent the "jewels in the crown" and only cover a minute percentage of the total area of our three counties. Protecting these sites is vital but we must spread the net wider if we are to ensure that future generations may also enjoy a rich environment.

Through a programme of education and publicity we aim to increase understanding and awareness of wildlife and the countryside. We hope to provide opportunities for people to enjoy wildlife both in the countryside and in the town and to encourage people to care for the natural world around them.

Education is a vital part of the Trust's work, involving both the young and the old in exploring and learning about the natural world. Our major resource of over 90 nature reserves exist not only for the conservation of species but also for education and enjoyment; the Trust recognises that in its publication "*Educational Opportunities on Nature Reserves*" which encourages schools and youth organisations to use 40 of our sites.

Some of our nature reserves are not of outstanding wildlife value but provide invaluable opportunities for education and recreation. The Henry Stephen/C. S.

Lewis reserve in Oxford has no rarities and is used throughout the year by schools and youth organisations; the children benefit greatly from this opportunity to explore a pond and a woodland so close to the city. The Blenheim Farm reserve in Charlbury is BBONT's first "community nature reserve", managed by and for the local community who use the area for a variety of recreational and educational activities.

The junior wing of BBONT is WATCH, the national environmental club for all young people who care about wildlife and who wish to play an active role in conservation. Local WATCH groups run by volunteers provide opportunities for young people to join in various activities and conservation projects, often held on BBONT nature reserves.

BBONT puts considerable resources into both education and publicity; staff are employed specifically to develop these important areas of the Trust's work. Membership and fund raising are also vital to the success of the Trust and are yet another aspect to our work.

Discovering secrets at a WATCH meeting.

VISITING NATURE RESERVES

We welcome visitors to the nature reserves described in this book; you may visit most of the reserves at any time but it is necessary to obtain a permit to visit those described in the "Permit only" section. We hope that you will enjoy our reserves and that you will return to visit them at different times of the year.

In order to help us to safeguard the nature reserves we ask that you follow the code of practice described below.

Code of practice when visiting reserves

1. Please do not remove any plants or animals from reserves.
2. Please keep to the paths wherever possible. Close all gates and avoid trespass or disturbance to adjacent property.
3. Take care to avoid disturbing birds or other animals especially during the Spring and early Summer when they are breeding.
4. Guard against fire risk; plantations, woodlands and heathlands are highly inflammable.
5. Do not leave litter; please take it away with you.
6. Sheep and cattle are often used to graze reserves for the benefit of the grassland plants; please avoid disturbing these animals.
7. Dogs are prohibited on some reserves and are only allowed into others on a lead.
8. BBONT accepts no responsibility for any loss, injury or damage, howsoever caused, which may be sustained while visiting the reserves described in this book.

Permit only reserves

Some of our reserves have restricted access and in order to visit these sites you must obtain a permit. There are several reasons for this; some sites are extremely susceptible to trampling or disturbance and in other cases the owner may wish to restrict the number of visitors to his or her land.

If you wish to visit a "permit only" reserve please send a stamped addressed envelope to the BBONT office requesting a permit and access details for the reserve that you wish to visit. We would also like to know when you intend to visit the reserve.

USING THIS BOOK

This book has been carefully planned for easy use; the reserves are listed alphabetically under each county, apart from the "permit only" reserves which are described in a separate section. At the end of each county section there are a few blank pages for notes and information about new reserves. The text and the map for each site are arranged for ease of reference.

Each reserve is described by its area, nearest towns, location and access, description and management. Months when wildflowers can be expected to be in bloom are indicated by the month numbers in brackets, eg; ragged-robin (5–7). Reserve maps show boundaries, where to park, where to enter the reserves, the main footpaths and habitats.

Grid references

The locations of the reserves are given as grid references on the relevant ordnance survey scale 1:50,000 Landranger series maps. The Trust has reserves located on the following maps:

No. 151 Stratford-upon-Avon
No. 152 Northampton and Milton Keynes
No. 163 Cheltenham and Cirencester
No. 164 Oxford
No. 165 Aylesbury and Leighton Buzzard
No. 174 Newbury and Wantage
No. 175 Reading and Windsor
No. 176 West London

The reserves can be located by using the six figure grid reference number given for each site; in case you are unfamiliar with this reference system it is explained below:

eg: SU 725 884

1. The first two numbers represent the vertical lines on the map within the square indicated by the letters.
2. The 4th and 5th numbers represent the horizontal position on your map.
3. Each vertical and horizontal line is numbered at the edge of the map. To use the reference find the intersection of the lines in 1. and 2. Imagine a letter L; the reserve will be in the square bounded by the L.
4. The 3rd and 6th numbers represent 1/10 marks between the whole numbers of each line and can be used to pinpoint a precise location within the square.
   eg: For SU 725 884 the point is 5/10 between 72 and 73 and 4/10 between 88 and 89.

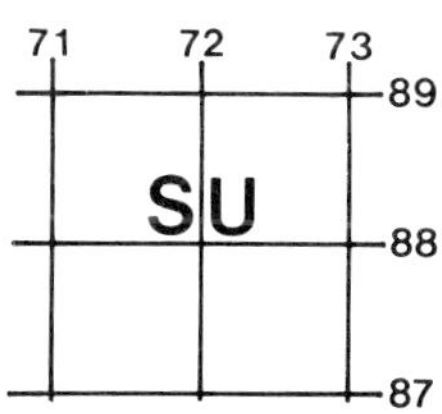

List of abbreviations used in this handbook

BTCV British Trust for Conservation Volunteers
NCC Nature Conservancy Council
RSNC Royal Society for Nature Conservation
SSSI Site of Special Scientific Interest (ie a site of national importance in terms of its wildlife or geology).

BBONT Nature Reserves February 1989

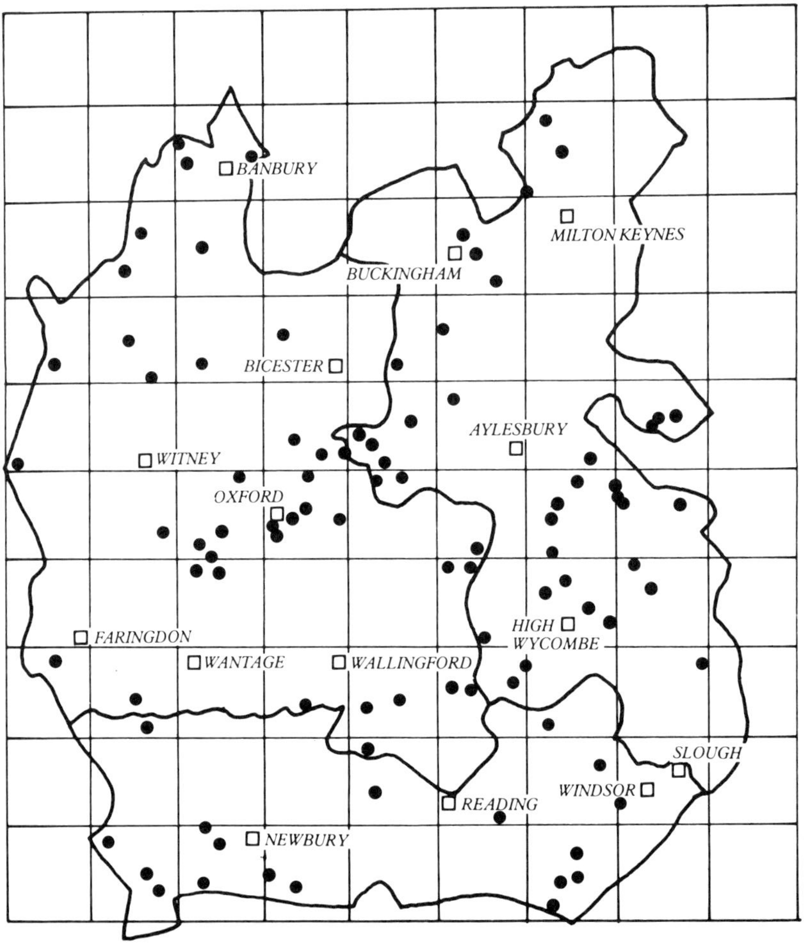

BERKSHIRE Reserves

Avery's Pightle
Baynes Reserve
Blackwater Reach Meadow
Bowdown Woods
Bray Pit
Broadmoor Bottom
Chawridge Bank
Hungerford Marsh
Hurley Chalk Pit
Inkpen Common
Inkpen Crocus Field
Moor Copse
Northerams Wood
Owlsmoor Bog and Heath
Seven Barrows
Sole Common Pond

AVERY'S PIGHTLE

OS sheet 174; SU 435651

Nearest town Newbury

An old flower-rich meadow, 1.3 hectares (3.18 acres) in area. This SSSI was bought by BBONT in 1985.

Location
To find this reserve, you need to locate Enborne, just to the west of Newbury. At the east end of the village, Church Lane leaves southwards towards Crockham Heath. After half a mile this road crosses a small stream, and the reserve is on the right hand side past (to the south of) the stream. Please park off the narrow road.

Access
Open to the public. Dogs are not allowed on the reserve when sheep are grazing in late Summer and early Autumn.

Description
'Pightle' (pronounced to rhyme with title) is a medieval word for a small enclosed meadow. Such fields were once a part of every village, but in recent times most of them have been drained, deep-ploughed and fertilised for high yield grasses or grain.

Traces of an ancient ridge and furrow system show that this field has not been ploughed for hundreds of years. The fact that herbicides and artificial fertilisers have not been applied is shown by the sheer variety of plant life: 113 species of flowering plants plus 24 wild grasses, sedges and rushes have been recorded. It is a mistake to think that the application of fertilisers will result in more flowers – what happens is the reverse, for the new supply of nutrient encourages the coarser growths to get ahead and swamp the other species!

16 plants normally confined to ancient grasslands are seen here. They are unevenly distributed across the field, reflecting differences in drainage and management. In Spring, on the drier, shorter-grazed area to the south, look for adder's-tongue, a small fern looking like a green echo of cuckoo pint with its small spathe and 'tongue'. Here pepper-saxifrage (flowering 6–8) and sneezewort (7–8) also grow, whereas dyer's greenweed (6–8), meadow thistle (6–8) and betony (6–9) are seen on the wetter ground to the north.

There are also wild orchids – the scrubby area near the stream in the north-east corner holds a colony of broad-leaved helleborine (7–9).

It is also a rich place for butterflies and birds. Butterflies seen here include large and small skippers, orange-tip, meadow brown, common blue, and green-veined white. Other insects abound.

Birds which have been noted include lapwing and blackcap, lesser whitethroat, yellowhammer and a variety of warblers attracted by the hedgerows. Some of these regularly nest here.

Management
In recent years before becoming a BBONT reserve it was grazed by ponies; now it is grazed by the BBONT sheep in late Summer and early Autumn.

AVERY'S PIGHTLE

OS sheet 174; SU 435651

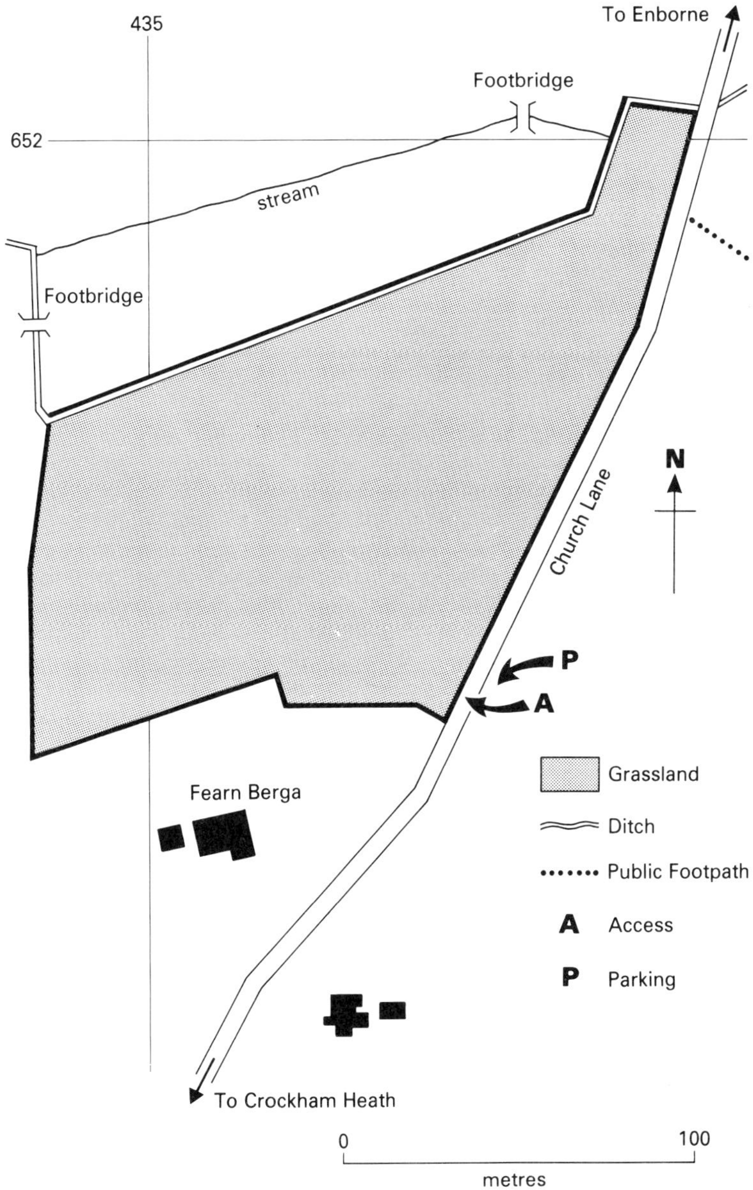

BAYNES RESERVE

OS sheet 174; SU 512652

Nearest town Newbury

An ancient woodland, 16 hectares (40 acres) in extent. Part of the Bowdown Woods and Chamberhouse Woods SSSI, BBONT has owned the freehold since 1983.

Location
From Thatcham just to the east of Newbury, take the (unclassified) road south past Thatcham station, over the River Kennet, and as the road rises to Crookham Common take Bury's Bank Road (a narrow right fork) which soon meets Greenham Common airfield. After half a mile, turn right down a gravel track (SU 511649) and pass Thatched Lodge to a small car park. The reserve is in front of you.

Access
Open to the public along the paths. Dogs must be kept on a lead.

Description
This is ancient woodland. It has been woodland for many centuries, perhaps since the return of trees at the end of the last Ice Age. However, this does not mean that it is wildwood untouched or unmanaged by man. No British woodland is!

The reserve covers two stream valleys converging to the north-east. Cutting down through layers of gravels and clays as they do, these valleys provide several distinct habitats. The higher, well-drained slopes with rather acid soil support birch and rowan with patches of oak and hazel and with honeysuckle (flowering 6–9, its red berries 8–9), foxglove (6–8), and several ferns; and bramble (5–9), bluebells (4–6) and bracken. Further down the slopes, the wetter and richer soils are marked by alder and plentiful hazel. Look here for the unusual 5-ways-facing flowers of town-hall-clock or moschatel (4–5), and for cuckoo pint or lords-and-ladies (flowering 4–5, with red berries 7–8) which is perhaps more familiar in hedgerows, and white wood-sorrel (4–5), folding its leaves down in rain. Springs emerge on the slopes creating 'wet flushes', marked in April and May by splashes of gold when the opposite-leaved golden-saxifrage is in flower.

These woodlands are alive with birdsong. The nightingale sings here from late April to late June (it likes dense low cover such as coppice or scrub to nest and will desert woods which grow too leggy).

Roe and muntjac deer are present, as is the dormouse. The clue to its presence might be that the honeysuckle is neatly stripped of bark 3–8 ft from the ground – this small animal uses the bark for its nest. The handsome white admiral and the majestic purple emperor butterflies are on the wing here in July and August. The visitor will see many species of dragonfly (15 in all have been recorded) – some of them breed in the streams but most fly in from nearby ponds.

Management
The old coppicing regime which regularly opens up a wood and creates diversity has been continued on a modest scale.

BAYNES RESERVE

OS sheet 174; SU 512652

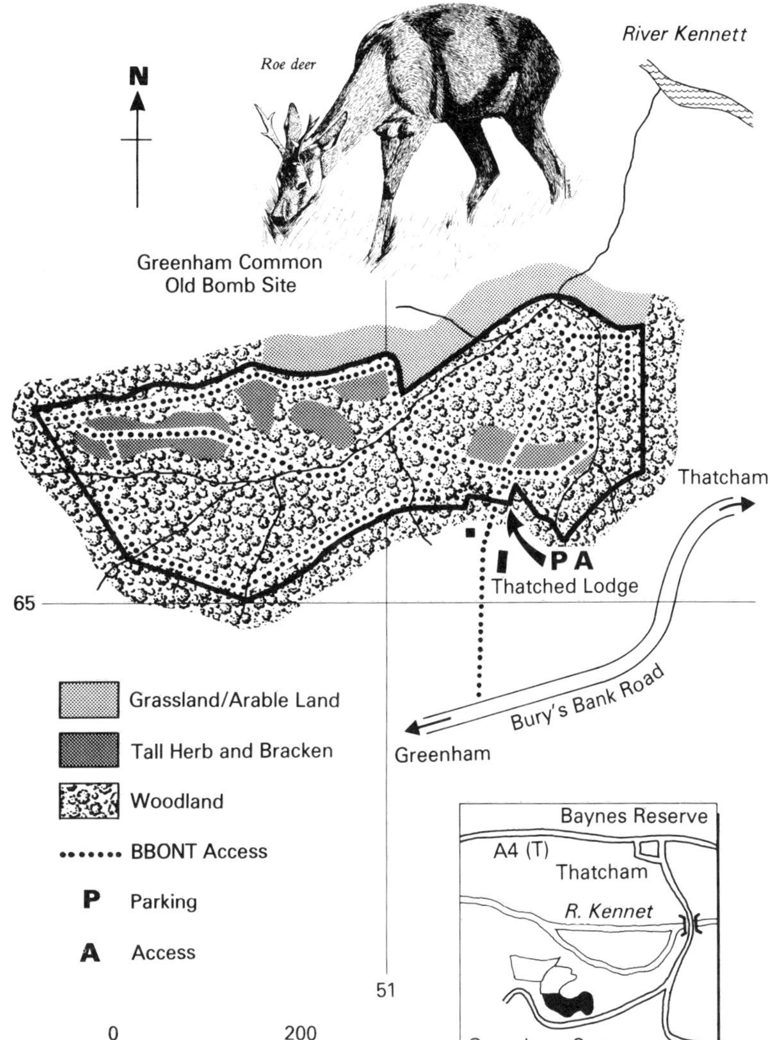

BLACKWATER REACH MEADOW

OS sheet 175; SU 843608

Nearest towns Bracknell, Sandhurst and Camberley

An 'unimproved' rough meadow of 1.2 hectares (3 acres). Part of the Blackwater Valley SSSI, it forms part of the Sandhurst Memorial Park. It has been managed by BBONT by agreement with Sandhurst Town Council since 1985.

Location
It lies within Sandhurst. Leaving Sandhurst on the A321 towards Camberley, you will come to the Town Council offices and playing fields on your right. Park here and walk south to the River Blackwater. The reserve is the last field before the river, beyond a small lake.

Access
Open to the public.

Description
A high quality, rough wet meadow, unimproved (that is, unchanged) by modern farming. Purple moor-grass, and betony (flowering 6–9) are seen; they are two of a number of plants which are seldom found outside unimproved meadows such as this, or reflect a long period of uninterrupted grassland management without the use of fertilisers or pesticides.

Over 100 species of wildflowers have so far been recorded. The list includes ragged-robin (flowering 5–6) and meadowsweet (6–9) growing in the marshier areas. The scent of the latter wafts across the reserve on still, humid Summer evenings. Among the other flowers to interest the visitor are cuckooflower or lady's smock (4–5), yellow iris (5–7), tormentil (6–9), water forget-me-not (5–8) and four species of buttercup, mainly in flower in June.

With such a variety of plants comes a variety of insect and other invertebrate life. The butterflies make the reserve a delight. One of the pleasures of the traditional countryside was to be able to walk through butterfly fields like this one, but such places are now rather hard to find.

Amongst the species here are orange-tip (which often lays its eggs on cuckooflower), green-veined white (which may choose the same species of plant) and common blue. This last is worth looking out for. It lays on various plants of the pea family and is widespread throughout Britain, but is not seen in large numbers in any one place.

The strongly flying large white and peacock butterflies have also been seen here, presumably breaking a journey to feed on the nectar of the flowers. Water measurer and great diving beetle have been recorded in the central ditch and grass snakes have been seen nearby.

The kingfisher is one of the birds recorded on the reserve.

Management
Areas are now cut in rotation in late Summer. A new footpath is planned along the riverside.

BLACKWATER REACH MEADOW

OS sheet 175; SU 843608

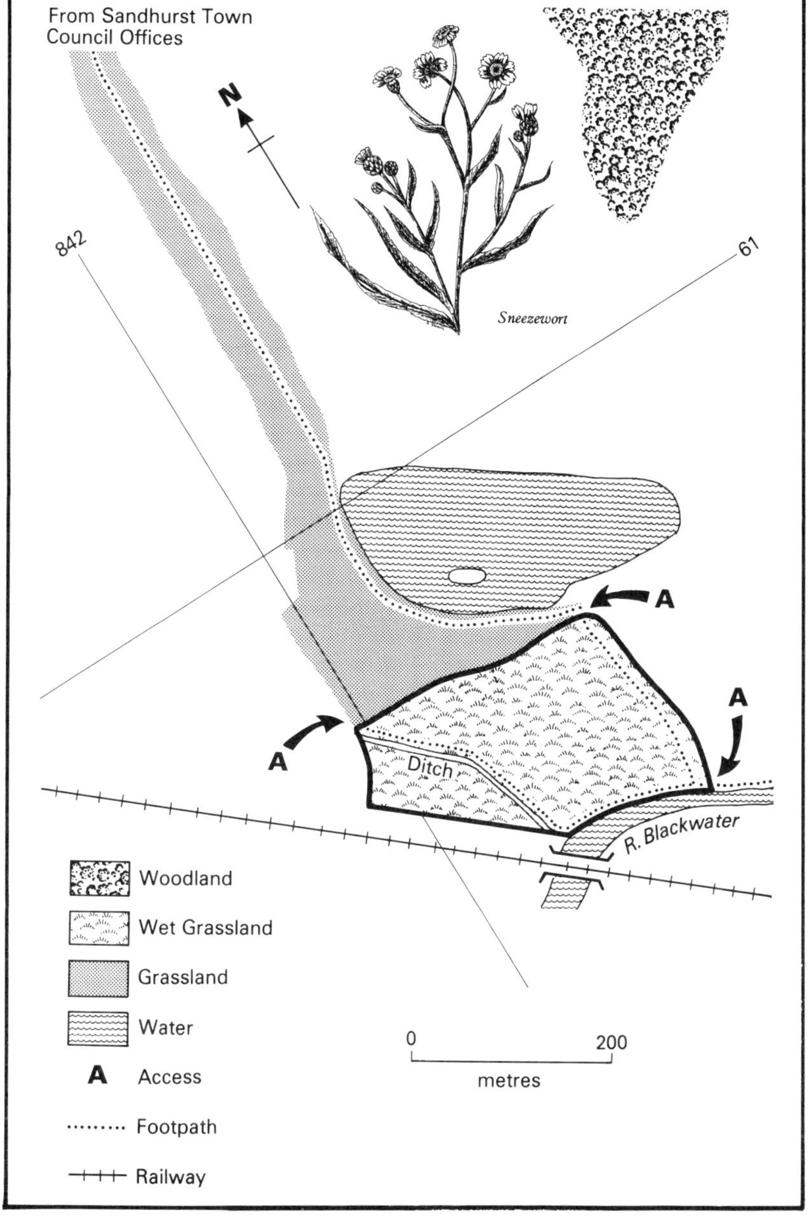

BOWDOWN WOODS

OS sheet 174; SU 504657

Nearest town Newbury

A varied reserve of 22 hectares (55 acres) of which 20 hectares (50 acres) lie within the Bowdown and Chamberhouse Woods SSSI. The reserve was bought by BBONT in 1984.

Location
Leave Newbury southwards on the A34. At the roundabout at the top of the hill turn left to Greenham village, and drive for about 1½ miles along Bury's Bank Road until you reach a small track on the left, with a sign to Newbury Trout Lakes. Take this track to a small car park on the right.

Access
Open to the public along the paths; dogs are not allowed.

Description
The reserve stretches for half a mile along a wooded slope where a gravelly plateau drops down to the River Kennet. London clay lies below the gravel and outcrops along the lower slopes. Spring-fed streams have incised numerous sharp valleys in the slope and there are also areas of seepage.

On this varied geological stage, vegetation creates its own scenery. There is dry heathland on the plateau and some patches of oak woodland with wild cherry and mountain ash (rowan) to be seen – the brilliant red berries of the latter are eagerly taken by birds in Autumn. There is some hazel and scatters of birch, and on wetter ground stands of willow and of alder which has been coppiced in the past. There is coppiced woodland of ash and field maple on the lower, clay areas. There are areas of grassland and two ponds. Solomon's-seal (flowering 5–6) is one of the many spring flowers to interest the visitor.

The wet ground makes tracking exciting. The spoors of fox, roe deer and muntjac deer can be expected. Roe deer are rather secretive, but there is a good chance of finding clues towards the rutting season (June–August): perhaps trees with frayed bark where antlers were scraped free of velvet and which they continue to anoint with scent as a territory marker. During rutting, the roe buck gives out throaty grunts. The muntjac – hardly graceful, but squat and rather pig-like – often gives a loud, unexpected dog-like bark from cover to surprise visitors.

Dormice also live on the reserve and slow-worms can be seen.

White admiral and silver-washed fritillary butterflies are present and also the purple hairstreak, though perhaps less frequently seen for it tends to keep to the upper branches of the oak trees. There are a good many moth species and an uncommon slug for the specialist.

Management
Rides and paths have been opened up, some areas have been coppiced, invasive sycamore has been controlled and the area of heather on the brow of the reserve has been extended.

BOWDOWN WOODS

OS sheet 174; SU 504657

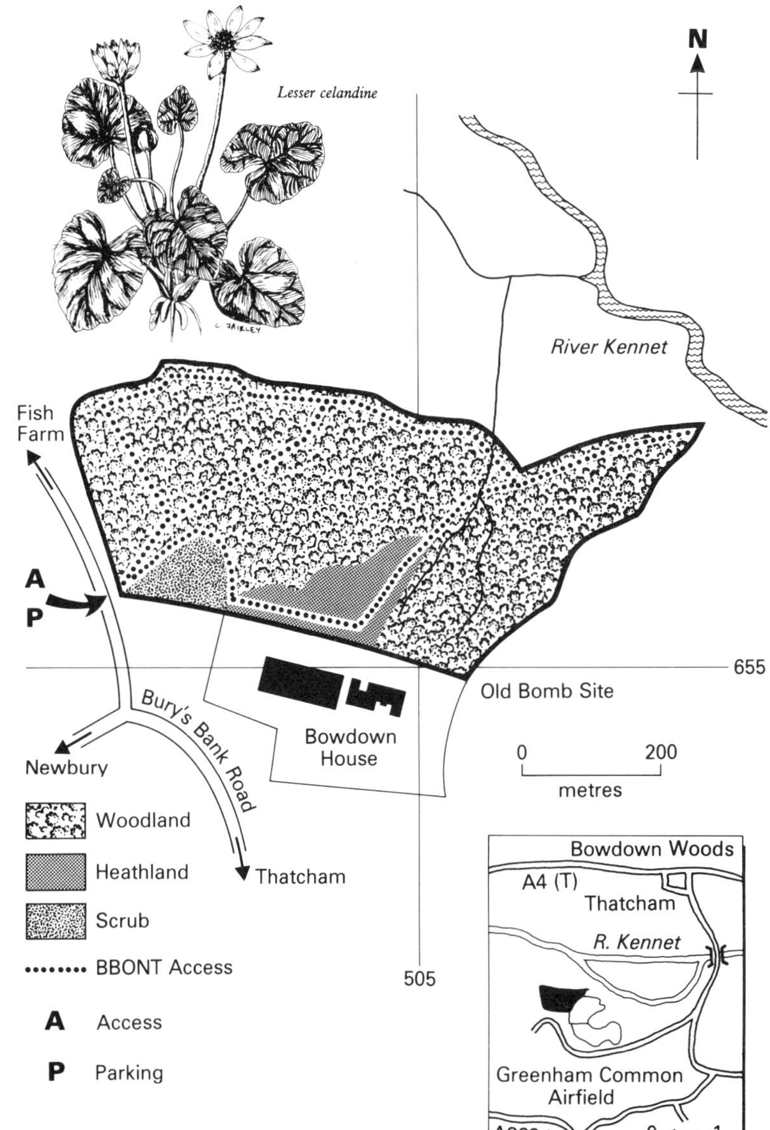

BRAY PIT

OS sheet 175; SU 904787

Nearest town Maidenhead

An old gravel working, now flooded, with varied surrounds. The reserve is 1.8 hectares (4.5 acres) in area and held on a peppercorn lease from the Summerleaze Gravel Co Ltd.

Location
Leave Maidenhead southwards on the A308 through Bray Wick and under the M4. Turn left immediately after the Brayfield Arms hotel into Monkey Island Lane. Drive down this road past the Bray Marina on your right and after a further 100 yards turn left (through a gate) into the water sports area. Continue past the Sailboard Clubhouse and park in the car park at the end.

Access
Open to the public.

Description
This a small wet sand and gravel working with a strip of woodland on its east side and grassy banks on the north-west and south-west sides. The M4 motorway forms the north-west boundary.

This reserve illustrates what can be created with modest resources.

It has a plant count of 50 species, but wildflowers have been planted in the meadow areas around the pit and now give a splendid display throughout the Summer. Species planted include oxeye daisy (flowering 5–9), various poppies (6–8), greater knapweed (6–8) and bird's-foot trefoil (6–9). Sneezewort (7–8) is now also to be found, a characteristic plant of old meadowland. Corn marigold (6–10) has been planted and also corncockle (6–8), a cornfield weed familiar to all in olden days which has been eliminated by the herbicides and seed cleaning techniques of modern farming. If ground with the corn it did taint it, but it is a very handsome flower for all that.

In other words, this reserve is something of a living museum.

Trees and shrubs have also been planted – wild cherry, wild pear and plum, alder and dogwood among their number.

There is varied butterfly life. On the wing in Spring are green-veined white and orange-tip. In Summer we see the meadow brown, once the emblem of long grass everywhere but it is now much less common in today's intensively farmed countryside. Those familiar garden visitors the peacock, red admiral and small tortoiseshell butterflies are seen here.

Bird life includes heron, Canada goose, mute swan and great crested grebe. Also of note are the common sandpiper, ringed plover and little ringed plover. The last is a Summer visitor, attracted by the shores of flooded gravel pits such as this.

Management
Planting of various (and sometimes unusual) vegetation as mentioned above. The muddy shallows to be left open to attract waders and as a loafing ground for duck.

BRAY PIT

OS sheet 175; SU 904787

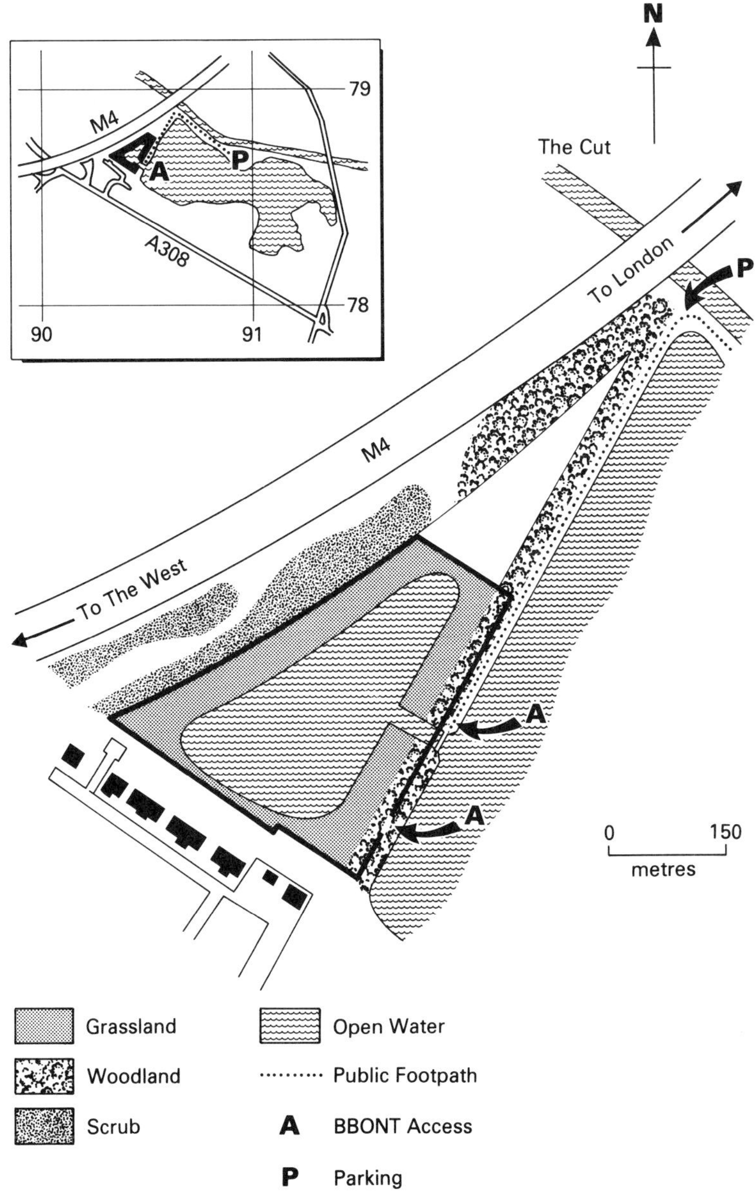

BROADMOOR BOTTOM

OS sheet 175; SU 855630

Nearest towns Bracknell, Sandhurst and Camberley

A mosaic of broadleaved and pine woodland, scrub, dry and wet heathland and ponds. The reserve is in two blocks, one south of Rackstraw Lane of 2.75 hectares (6.6 acres) and one north of 8.03 hectares (19.3 acres). A planned by-pass will divide the latter. Most of the reserve lies within the Broadmoor-Bagshot Woods and Heaths SSSI, and it is managed by BBONT in agreement with Broadmoor Hospital.

Location

From the centre of Sandhurst take the A321 east to the first roundabout and turn left. Cross two roundabouts until open land appears to the left. Drive for ⅓ of a mile and then park on the left where Owlsmoor Road crosses en route for Broadmoor Hospital. The map shows the access points.

Access

The smaller block is open to the public. The larger block is fenced and entry is by permit, obtainable from the BBONT Office in Oxford.

Description

Heathland must be treasured today: 80% of it has now been lost over the country as a whole. This relic, crossed as it is by pylons and edged by housing sprawl, is a last refuge in the neighbourhood for lizards, for beautiful orchids and the changing colours of the heather which delight many visitors. It is valued in many different ways.

Heathland came into existence when woodland was cleared long ago in late Stone Age and Bronze Age times; lack of land care, overgrazing of the poor soil and fires kept it open. The heathland here became part of the Royal Windsor Forest, a hunting domain, and grazing and cutting of heather and gorse by those with local common rights kept it open for the chase. But then early in the last century, field enclosure bit into the open heath and there was much planting of Scots pine. Grazing and cutting declined allowing trees to invade (the wooded areas were open before 1871). Then housing sprawl began to take its bite. By 1931, Broadmoor Bottom was one of the last fragments of the once extensive Bagshot Heath; in 1971 new housing to the south eroded it further.

Thus, of the range of closely interlinked habitats to be seen on the reserve, the open heathland and valley bog areas are of particular interest as relics of former times. In the valley bog grow bog asphodel (flowering 7–8) and bog myrtle (catkins 4–5), together with the 'insect eating' round-leaved sundew (6–8). Heather occupies drier soil.

There are many dragonflies, and one butterfly of note is the silver-studded blue. The handsome day-flying emperor moth might also be seen. Woodlark and stonechat are amongst the birds recorded as nesting here.

Management

The open heath must be kept clear of invasive scrub and trees. The woodland pond will be opened up to allow more light to reach the water.

BROADMOOR BOTTOM

OS sheet 175; SU 855630

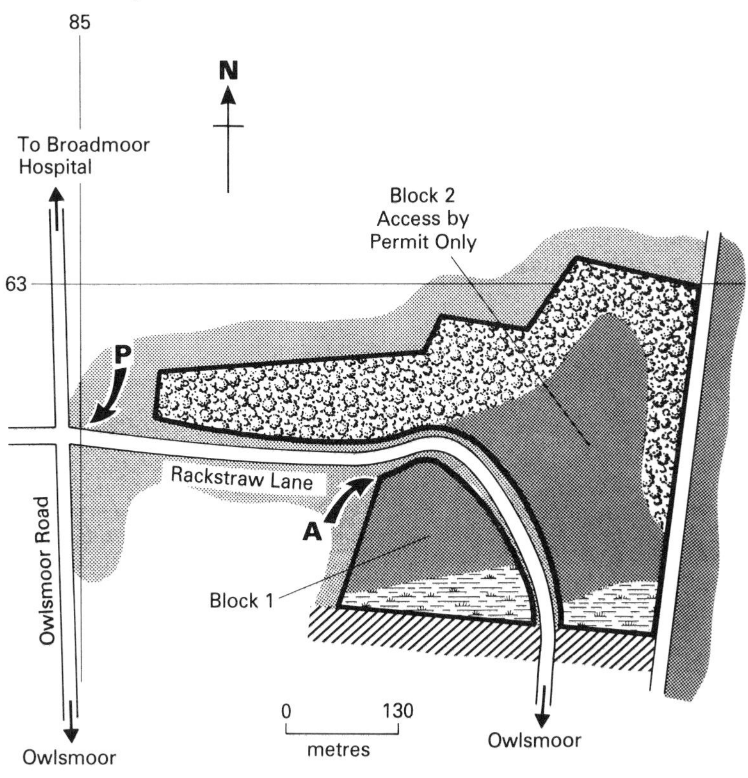

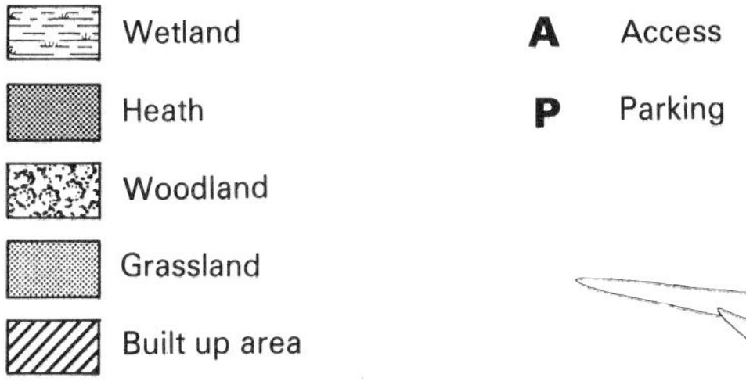

Silver-studded blue on cross-leaved heath

CHAWRIDGE BANK

OS sheet 175; SU 894738

Nearest towns Bracknell and Maidenhead

An area of unimproved grassland along the banks of a small brook, 3.2 hectares (8 acres) in area. An SSSI, BBONT bought the freehold in 1981.

Location
From Bracknell take the A3095 north to its junction with the A330 at Hawthorn Hill and turn right. Take the first turn to the left and proceed to a group of houses at SU 890736. Parking is difficult; please park off the road without blocking private entrances. Walk north along the bridleway (Hogoak Lane) to a gate on the right leading to the reserve, which lies on the bank across the Chawridge Bourne.

Access
Open to the public. No dogs or horses allowed.

Description
Although hawthorn and blackthorn scrub has invaded the northern end, much open grassland remains. This has escaped modern ploughing and agricultural improvement and many plant indicators of old grassland survive, such as adder's-tongue and pepper-saxifrage (flowering 6–8) typical of neutral to acid soils (here London clay). Many of these meadow species are now rather rare throughout BBONT's three counties (and indeed Britain as a whole). There are also cowslips (4–5), hairy violet (3–5), dyer's greenweed (6–8) and devils'-bit scabious (6–10).

Amongst the wild grasses are meadow foxtail, Yorkshire-fog, tufted hair-grass and red fescue. The last is found on the many large anthills, which are themselves another indicator of ancient unploughed ground.

An important feature of the reserve is the parish boundary hedge on the east bank of the stream, which contains field maple and oak pollards. Such hedges are often old and as a result contain a rich variety of woody species.

The reserve is excellent for butterflies. The dingy and grizzled skippers are on the wing in May and June. They are moth-like butterflies with a quick darting flight. The common blue is also seen here.

Harvest mice have been recorded. These charming, tiny creatures build nursery nests attached to the stems of tall grasses, and in the past they also built them in the cornfields where they were found at harvest time. The frequent passes of the spray in modern farming have made the harvest field a lethal place for much wildlife. They must have been very familiar, yet Gilbert White, the patron saint of British naturalists, describes in 'The Natural History of Selborne' how in 1767 he found some mice still 'non-descript' – undescribed in the scientific literature of the day. He popped them in a jar of brandy to preserve them! They were harvest mice.

Management
A flock of sheep controlled by the Reserve Manager maintains the interest of the grassland.

CHAWRIDGE BANK

OS sheet 175; SU 894738

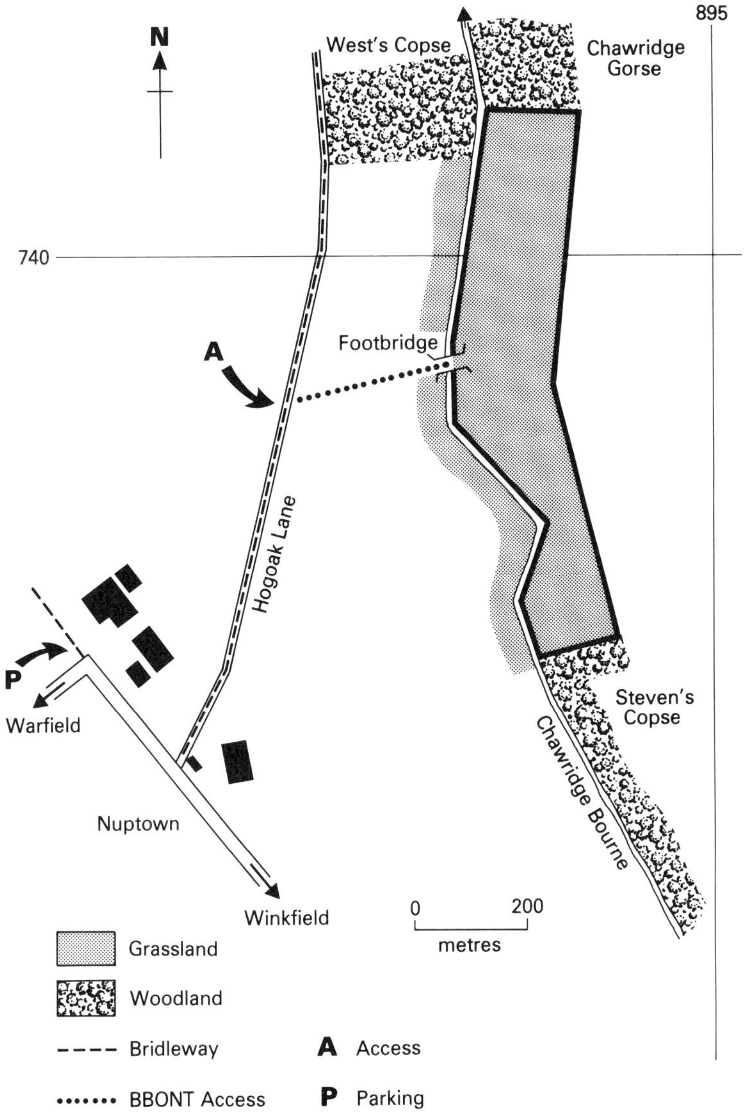

HUNGERFORD MARSH

OS sheet 174; SU 333687

Nearest town Hungerford

Unimproved rough grazing and reed bed, 11.2 hectares (28 acres) in area. Both the reserve and the adjoining Freeman's Marsh form part of an SSSI. The reserve is managed by BBONT by agreement with the owner.

Location
On the western edge of Hungerford, lying alongside the Kennet and Avon Canal and with the River Dun winding through it. Access can be gained via the public footpath which crosses the canal via a swing bridge near the church. A line of willows and bushes on the western boundary separates it from Freeman's Marsh, managed by the Town and Manor of Hungerford.

Access
Open to the public (but please keep to the footpath).

Description
The River Dun flows across peaty ground. In the distant past it was wetland, but today the reserve consists of rough grazing with some remaining areas of reed bed. There is some alder woodland.

Colourful wetland wildflowers are still seen, including yellow iris (flowering 5–7) and ragged-robin (5–6). When in flower (3–5) the handsome yellow marsh-marigold or kingcup can be seen from afar. It was familiar to all in our grandparents' day, but with the widescale field drainage and 'improvement' of recent times we can no longer take it for granted.

Several species of birds nest along the river banks and on the marshy meadow. Reed and sedge warblers nest in the area of the reed beds. The former expertly weaves a deep nest slung between the reed stems, while the latter nests in other vegetation near the edge. They may be more frequently heard than seen among this dense cover; both species have songs with some liquid and some grating notes.

Snipe used to nest regularly on the reserve, and it is hoped that the current management will encourage them to do so in future. They are not as familiar as they once were and the reserve will be well worth visiting just to witness their courtship display. In courtship, the cock bird climbs quite high with rapidly beating wings, then dives with its outer tail feathers spread. These vibrate in the slipstream to create an odd, resonant bleating note or 'drumming' as it is called.

There are records of 120 different bird species over the last ten years or so, including heron, kingfisher, yellow wagtail, water rail and grasshopper warbler. This last may nest – and if so its unusual song will be heard, a repetitive high pitched machine-like whirr delivered from deep cover. In Winter, small numbers of waterfowl are regularly seen and siskins are frequent visitors to the alders.

Management
The area to the south of the river is Winter-grazed by cattle; that to the north is not grazed to allow the reed bed to expand. Some hawthorn is being removed. The willows are pollarded at intervals.

HUNGERFORD MARSH

OS sheet 174; SU 333687

Marsh-marigold

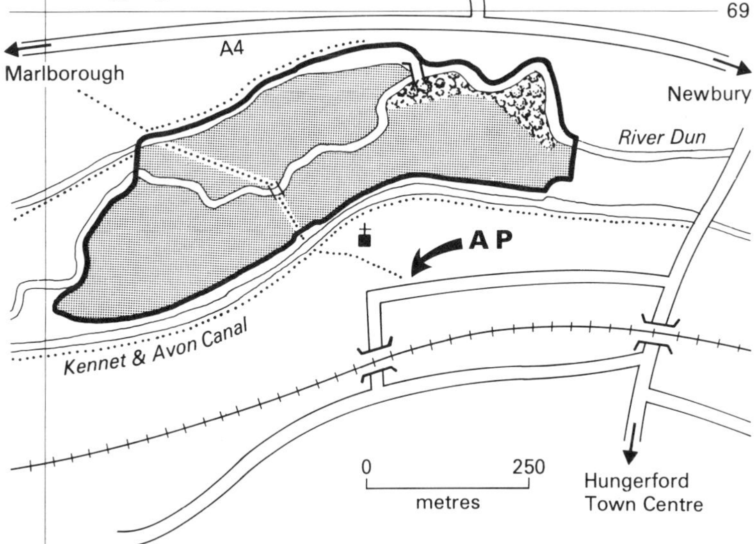

Snipe

HURLEY CHALK PIT

OS sheet 175; SU 813820

Nearest town Henley-on-Thames

A small beechwood, chalk pit and an area of chalk grassland, 1.4 hectares (3.5 acres) in extent. The freehold was bought by BBONT in 1964. It was BBONT's first acquisition by purchase.

Location
The reserve is reached on the A423 from Henley to Maidenhead. Just to the west of the Black Boy Inn (SU 812830), about 3¾ miles from Henley, there is a layby on the north side. Park here and take the bridleway leading south from opposite the Black Boy. The reserve is a ¾ mile walk away.

Access
Open to the public.

Description
In spite of its name, this reserve only consists in part of a chalk pit, which lies at the south-west corner. This has chalk cliffs and scree. Flints and fossils are on open view, but please do not collect the fossils.

Most of the reserve – around 2 acres of the total area – is beechwood growing on no-man's-land lying between what were once two highways (today's bridleways). Many of the beech have been pollarded and there is also coppicing. The ground is largely bare here and the wood is rather empty of insect life. The fungi may be interesting, however: there were reports of the unusual earth-star fungus growing here in the past.

Close by the pit is a small area of open chalk grassland which is quite the opposite. Here over 100 plant species have been recorded.

The plants include a variety of wild grasses, most easily recognised when they are in flower. Quaking-grass (flowering 6–8) is typical of chalk grassland.

There are small numbers of wild orchids. Common spotted- (6–8), fragrant (6–7) and bee orchid (6–7) might be seen. The local Chiltern gentian also flowers here (8–9), bearing tubular, rather purplish flowers.

There is also a small area of mixed scrub in the south-west corner and some hazel coppice in the south-east corner.

This variety of plant life is reflected in the number of butterflies seen here – no fewer than 20 different species have been recorded. They will either lay their eggs on the wildflowers or visit them for nectar. The total includes the marbled white (which in spite of its name and patterning is one of the Brown family of butterflies).

All in all, the reserve is interesting and varied, as befits BBONT's first purchase.

Management
Scrub is regularly cut back to keep the chalk grassland open. The grassland is cut in the Autumn. There is coppicing of hazel and beech.

HURLEY CHALK PIT

OS sheet 175; SU 813820

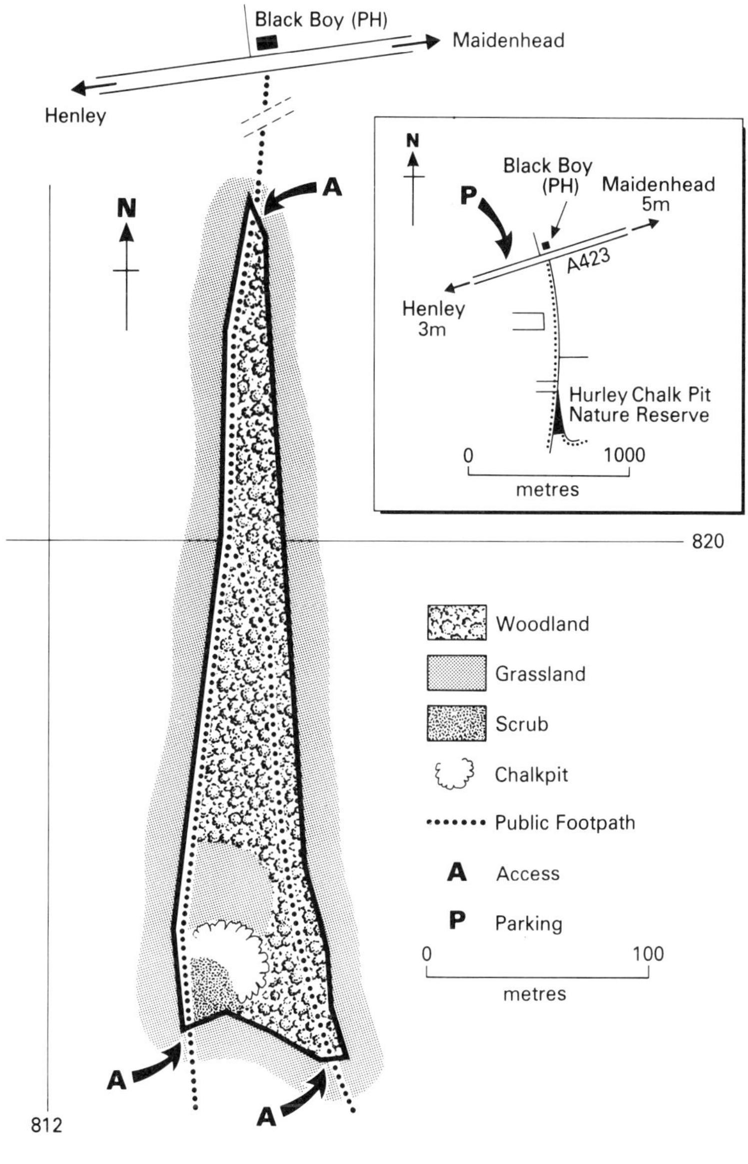

INKPEN COMMON

OS sheet 174; SU 382641

Nearest town Hungerford

Heathland (with bog) and areas of woodland, 10.4 hectares (26 acres) in extent. An SSSI, the reserve is on lease to BBONT.

Location
Leave the A4 3½ miles east of Hungerford, on the road leading south to Kintbury; take the second turn to the left in Kintbury village, following the road to East and West Woodhay for about 1¼ miles. Bear right at Rooksnest farm, signposted Inkpen Common. You will there find the main part of the reserve on the left opposite a row of small houses. Cars may be parked on the verge on the left of the road.

Access
Open to the public. Horse riding is restricted to the bridleway along the south-east boundary of the reserve.

Description
Acid soil such as we have here encourages heathland plants. The reserve is the remnant of the former Inkpen Great Common, and in the past it was kept open because the local people had rights of common to graze their livestock, collect firewood, heather and bracken (for bedding) and gorse (for firing bread ovens). Without this cropping, gorse and birch and other trees would quickly become established.

The reserve is in two parts as the plan shows, and the smaller south-western area is now a small wood of naturally regenerated oak and birch on an abandoned area of heath. Otherwise, an area of damp heathland survives, a 'wilderness' of a kind now very scarce in Berkshire. But even here there are patches of woodland growing at the expense of the heather and gorse. Although in full heavy scented bloom in Spring, common gorse always bears a flower or two ("when gorse is out of bloom, kissing is out of fashion", as they used to say). Dwarf gorse with spines too weak to prick is also present (flowering 7–9).

The heathland carries purple moor-grass (in flower rather late in the year, 8–9) and other grasses, and heather. Or, to be more precise, heather (7–10), bell heather (7–9) and cross-leaved heath (6–9). They can be distinguished not only by the set of their small leaves, but by the fact that the first two occupy drier ground and cross-leaved heath wetter ground. Lousewort (4–7) flowers abundantly in some places, and the uncommon pale dog-violet can be seen (5–6).

The reserve includes a small valley bog, fed by seepage. The most dramatic flower here is bog asphodel with glorious yellow flowers (7–8); it is equally striking later, when both seed pods and leaves become orange. Meadow thistle (6–7) and petty whin (4–7) also mark parts of this wetter area.

Management
The invasive gorse and birch need to be controlled to allow the heathers to flourish.

INKPEN COMMON

OS sheet 174; SU 382641

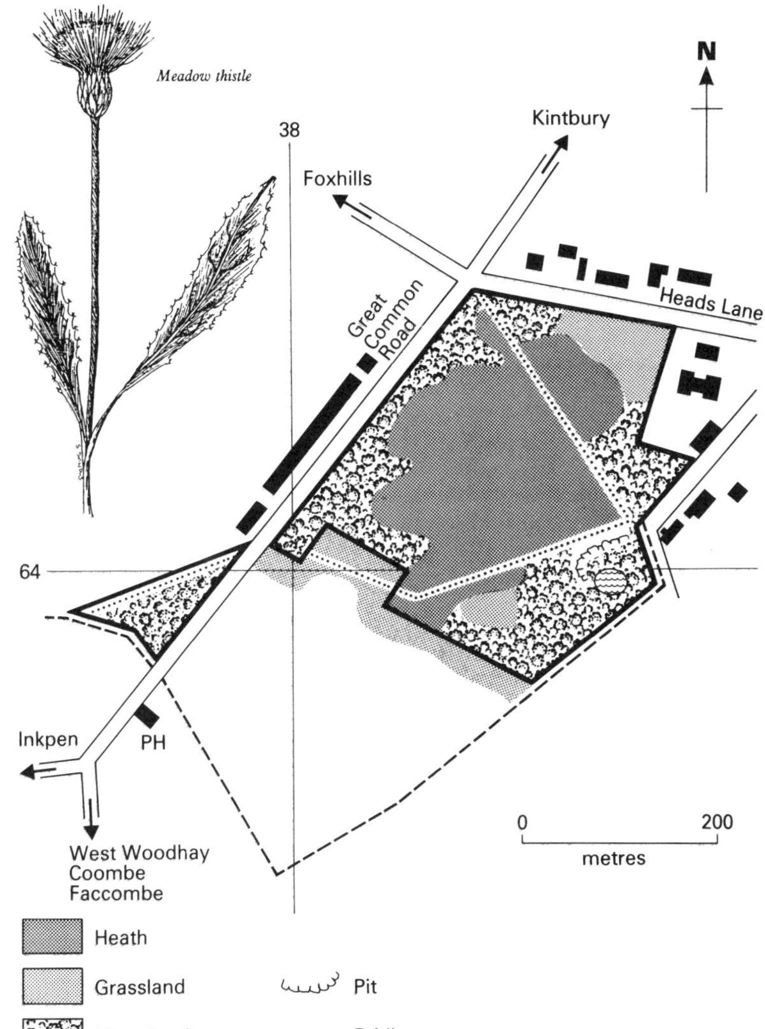

INKPEN CROCUS FIELD

OS sheet 174; SU 370640

Nearest town Hungerford

A thickly-hedged, flower-rich meadow site, notable for its wild crocus. 3.2 hectares (8 acres) in area, it is an SSSI, bought by BBONT in 1986.

Location

From the A4, 3½ miles east of Hungerford, take the road leading south to Kintbury. Take the second left in Kintbury village, for Inkpen. After a mile turn right at cross roads, and fork left after half a mile. After a further half mile turn right onto a narrow concrete track (Pottery Lane). After 200 yards there is a grass track on the left with a gate at the end leading into Crocus Field.

Access

Open to the public. Dogs must be kept on a lead.

Description

This reserve boasts the largest population of spring crocus in Britain, thousands of the purple or white-purple flowers being seen in March–April. The flowers appear with the leaves.

The spring crocus is not a native species, and tradition has it that the plant was originally brought back by the Crusaders. If so, the reason is now unknown, for it is not the crocus which yielded saffron for flavouring and colouring sweetmeats in olden days. How it came to be growing here is even more of a puzzle – the duller explanation would be that it came with garden rubbish dumped long ago on the site – parts have been dug for clay at some time in the past and these diggings may have been used as a local waste tip. However, written records show that the flowers have been here since 1800 and farmers delayed grazing to protect them!

Crocuses apart, these two fields separated by a small, Spring-fed stream are a fine example of flower-rich meadowland on clay soil. It is interesting to see how change in the character of the undisturbed soil is reflected by the plants found. The northern meadow is rich in wild grasses and wildflowers (and has more crocus) while the southern has ranker grass, fewer flowering plants and rushes noticeable in the wetter parts. Some 149 plant species have so far been recorded on the reserve, some of which are found only in 'unimproved' grassland such as this. One of interest is pignut – a rather shorter relative of cow parsley. It grows from underground tubers and is seen in flower in May and June. These tubers are the 'nuts in May' gathered in the children's nursery rhyme.

As is to be expected, with a galaxy of wildflowers such as this we also find a variety of butterflies including small and common blues.

Management

In the past there was spasmodic grazing – horses were kept at times. It is now grazed by BBONT's Beulah sheep which are put on after the crocuses have finished flowering.

INKPEN CROCUS FIELD

OS sheet 174; SU 370640

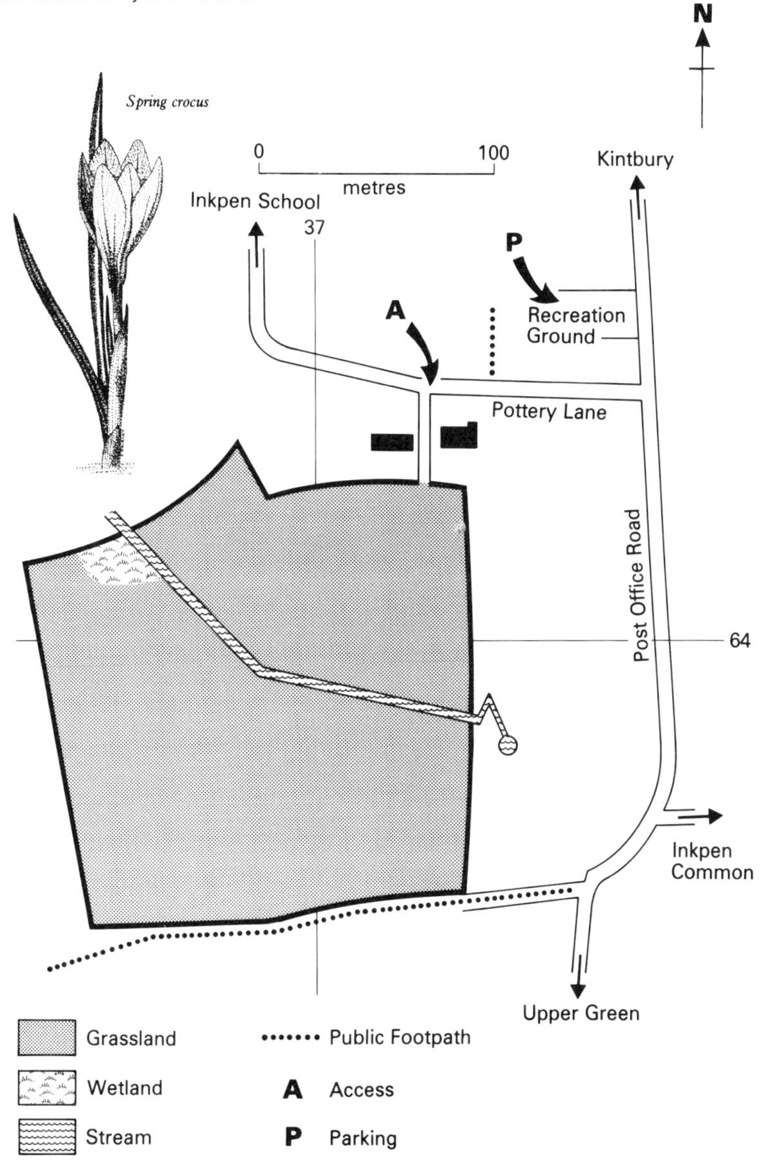

MOOR COPSE

OS sheet 175; SU 635742

Nearest town Reading

Varied woodland, 26.8 hectares (67 acres) in area and the major part of an SSSI, BBONT were left freehold ownership in 1975 and acquired Bartons Copse in 1988. The length of the River Pang running through the reserve was bought by BBONT in 1975.

Location
Leave Pangbourne southwards on the A340. Just past Tidmarsh, as you rise to cross the M4 turn left into a layby with parking. The reserve is on the left.

Access
Open to the public.

Description
The diversity of the reserve is only revealed on walking deep into it. Hogmoor Copse, the first area to greet the visitor, is a wet wood on peat and gravel, with willows and a few oak and ash trees. Although once coppiced, it is now a tangled, jungly place. Drainage ditches have been dug across it in the past, the reason for which is not clear.

Park Wood on the other hand is a type of ancient landscape very familiar to our ancestors, although not much like film versions of the greenwood. This ancient woodland consists of areas of coppiced alder and hazel, overshadowed by giant 'standard' oak trees. The coppicing opens up the wood to give cathedral-like vistas. It also, by opening the woodland floor to light, encourages the primroses (flowering 3–6) and the bluebells (4–6) which here make a dramatic show. There is also a small stand of Norway spruce (Christmas tree) in Park Wood.

The 5 acre hay field makes an interesting contrast, while beyond it lies another coppiced wood, Moor Copse itself. This is an old ash coppice with widely spaced oak and ash tree standards. Early-purple orchid (4–6), one of the indicator plants of old woodland, is present.

Large numbers of moths have been recorded (over 250 species). The handsome white admiral is seen on the wing in July and early August; this butterfly lays its eggs on the abundant honeysuckle. Beetles and other invertebrates flourish in these woodlands, reflecting the large quantity of fallen and cut wood left to rot for their benefit.

Muntjac, roe deer and foxes are regularly seen. Grass snakes are found basking in the sunshine in the newly cut coppice plots. A variety of woodland birds are seen here, including the sparrowhawk and tawny owl.

The River Pang itself must not be forgotten. It is unpolluted and holds a good variety of insects and fishes. There can be a heavy mayfly hatch from this river in May and June.

Management
The rides are kept open and much of the woodland is in the process of being coppiced in rotation.

MOOR COPSE

OS sheet 175; SU 635742

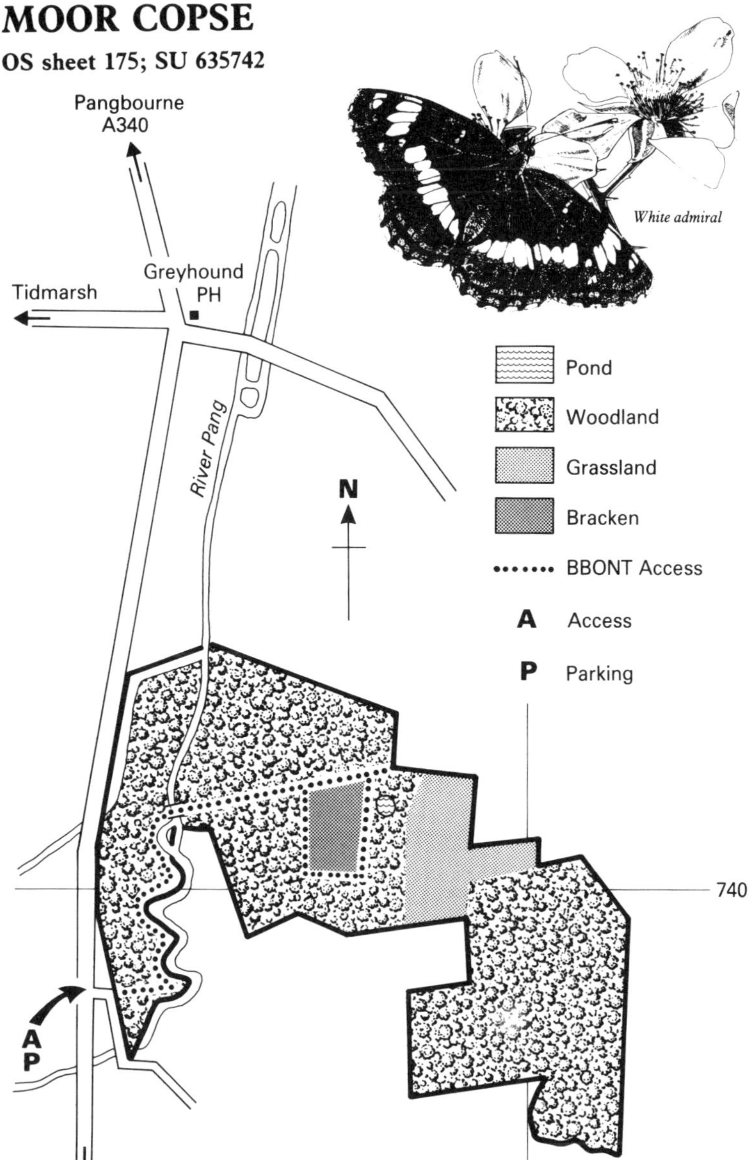

White admiral

NORTHERAMS WOOD

OS sheet 175; SU 856682

Nearest town Bracknell

A 3 hectare (7.5 acre) relic of an old woodland. It was leased to BBONT by Bracknell District Council, and declared a nature reserve in 1980.

Location
This is an urban reserve, within Bracknell, adjoining the A3095, the main road from Bracknell to Crowthorne. The plan explains the route from this main road (which is also known as Mill Lane). At the roundabout take the road to the west signposted Great Hollands, and follow it round to the right until you reach a car park on the right hand side. Park here and walk to the reserve, which is across the playing fields.

Access
Open to the public.

Description
This wood was saved by the action of local residents. A remnant of a much larger area of woodland, now largely disappeared under factories, it is valuable for being a wildlife-rich site within a built-up area.

It is mixed woodland with small areas of open damp meadowland and some scrub. Within the woodland, three kinds of habitat can be recognised – oak scrub, hazel coppice with birch, and a mixture of hazel, birch, sycamore, sweet chestnut, alder, aspen, holly, elder, wild cherry and hawthorn. It is surprisingly diverse for its size. More than 100 plant species have been recorded here, and there is a good show of wildflowers in Spring including bluebell (flowering 4–6), and yellow archangel (5–6).

18 different species of butterflies have been recorded including the white admiral and the brimstone. This last, the familiar yellow butterfly of Spring, perhaps gave the 'butterfly' its name. The females (which are in fact not yellow but white) search for buckthorn bushes on which to lay their eggs. The other species are more common in the open areas of the wood, however. The small copper, common blue, meadow brown, small heath and other browns are all linked with grassy areas. The painted lady has also been recorded here, a migrant butterfly arriving from North Africa and Europe.

The birds recorded as nesting include blackcap and garden and willow warblers. Woodpeckers have been heard drumming.

Management
The hazel in the woods is coppiced on a 15–20 year rotation; this opens up the woodland floor to light and encourages the wildflowers (and butterflies) and gives a more varied habitat for birds. Invasive sycamore is gradually being removed. The grassy areas are now mown each year to increase the diversity of the plant life (mowing or grazing regimes tend to keep ranker growths at bay and encourage the slighter plants) and the insects and other invertebrates which depend on them. A small pond has been dug.

NORTHERAMS WOOD

OS sheet 175; SU 856682

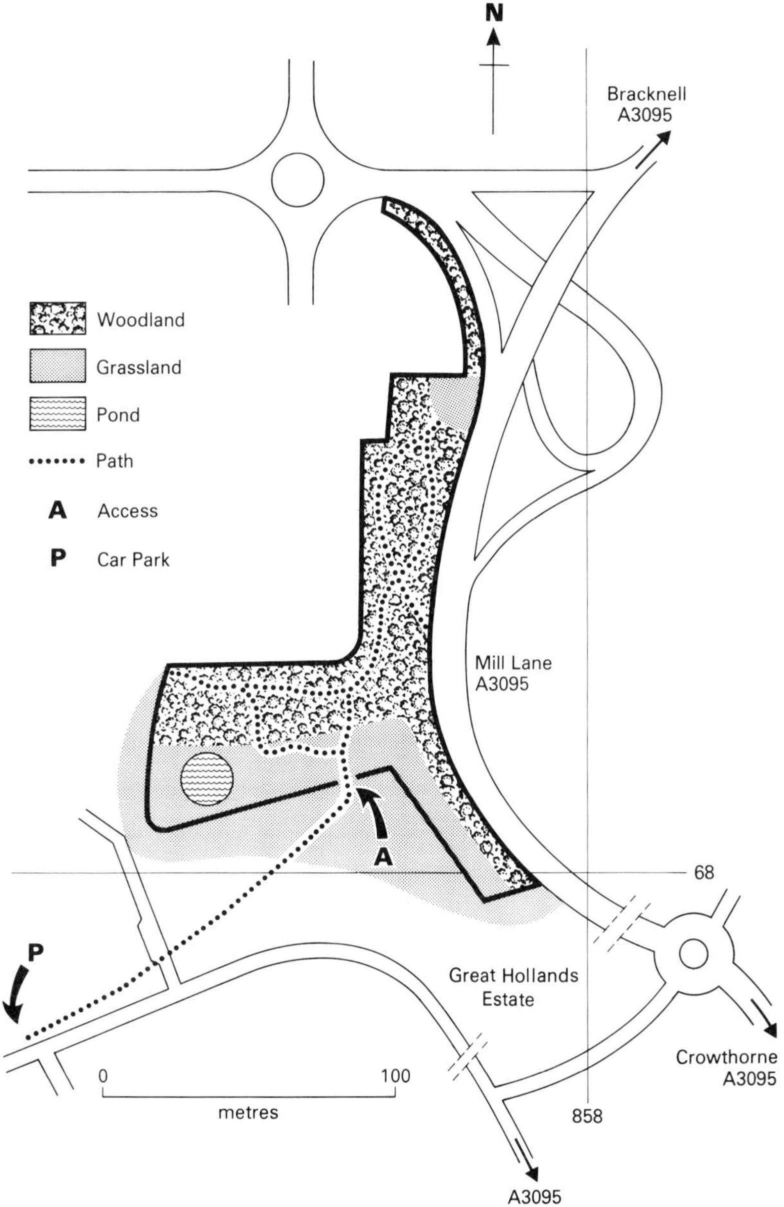

OWLSMOOR BOG and HEATH

OS sheet 175; SU 845632

Nearest towns Bracknell, Sandhurst and Camberley

A mixture of heathland, bog and woodland, 26 hectares (64 acres) in area. It is part of the Sandhurst to Owlsmoor Bogs and Heaths SSSI and is owned by BBONT, having been bought in two lots in 1985 and 1986.

Location
From the centre of Sandhurst take the A321 east (towards Camberley) to the roundabout and turn left. Cross two roundabouts until the heathland appears on the left. Use the District Council car park just south of Crowthorne at SU 838631 and walk along the bridleway just to the north, which forms the southern boundary for most of the reserve. Alternatively park where the Owlsmoor Road crosses en route for Broadmoor Hospital. The reserve is on the left and extends towards the A3095 between Sandhurst and Crowthorne. (See also BROADMOOR BOTTOM reserve). Edgbarrow Woods alongside (also within the SSSI) is managed by Bracknell Forest Borough Council.

Access
Open to the public; dogs must be kept under control and horses are allowed along the bridleway only.

Description
Together with the Broadmoor to Bagshot Woods and Heaths SSSI (part of which is also a BBONT reserve), the Sandhurst to Owlsmoor Bogs and Heaths SSSI forms 60% of all the open heath and bog remaining in Berkshire.

The reserve lies on sandy soils for the most part, acid and infertile, while a thin layer of peat has formed in the low lying areas. It is in these peaty areas that the bog is found. The agent in the creation of a bog is bog moss or sphagnum, its floppy stems creating a living waterlogged sponge. In hilly areas, these bog mosses can more or less survive on the nutrients brought by rain, and testify to our damp 'atlantic' climate. Round-leaved sundew (flowering 6–8) grows on the bog; small insects become trapped on their sticky hairs and are digested for their minerals.

The reserve has three species of heather and some bilberry (4–6) growing on drier ground. Such areas offer living ground to lizards, slow-worms and adders. These last are most likely to be seen sunning themselves after emerging from hibernation in Spring. They are not really dangerous, but shy, and likely to be well hidden by the time you reach the spot. The emperor moth is also seen, a handsome relative of the silk moth, quickly skimming over the heather tussocks in the sunshine.

Other parts of the reserve carry pine and birch woodland with some interesting fungi in Autumn.

Birds recorded nesting in the heathy areas include tree pipit, nightjar and stonechat. Many others nest in the woodland.

Management
The heathland is gradually being cleared of scrub and trees.

OWLSMOOR BOG and HEATH

OS sheet 175; SU 845632

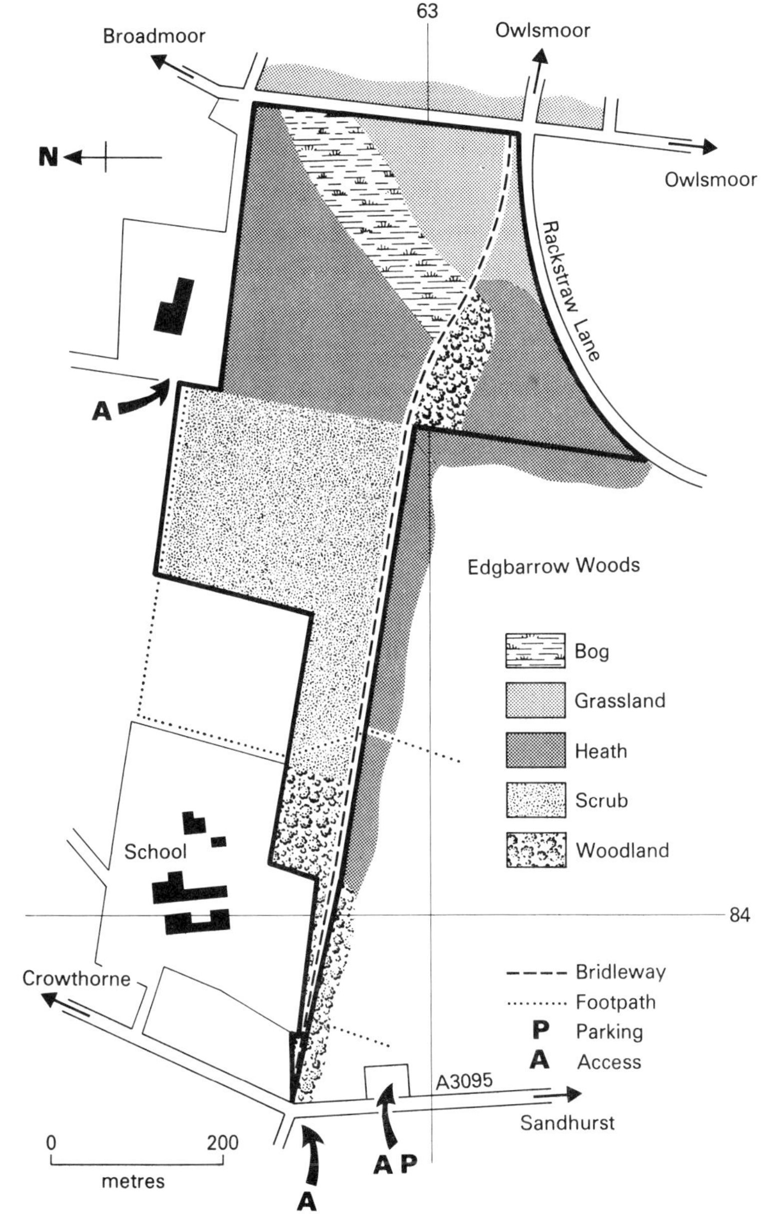

SEVEN BARROWS

OS sheet 174; SU 329828

Nearest towns Wantage and Lambourn

5.3 hectares (13 acres) of ungrazed chalk grassland encompassing a number of prehistoric burial mounds. It is a site of national importance, both for its archaeological interest and as a nature reserve. It is an SSSI and managed by BBONT in agreement with the owner.

Location
The directions start from Lambourn, to the south west of Wantage. Leave Lambourn northwards on the B4001 for 1½ miles to Mile End and then fork left for a mile. As the plan shows, the reserve lies along the north side of the road, between it and a bridleway. Cars should be parked off the road at the southern corner and not driven across the grass.

Access
Open to the public.

Description
This is an ancient and atmospheric site. There are in fact over 30 burial mounds of various kinds in and around the reserve, mostly dating from the Bronze Age. They have made it physically difficult to use the land for anything but grazing. It has never been ploughed.

The reserve is a precious fragment of the flower-rich grassland which until comparatively recently was widespread over the chalk of these Berkshire Downs. Over 150 different plant species have been recorded including the horseshoe vetch (flowering 5–7), and common rock-rose (5–9), the delicate blue harebell (7–9) nodding in the breezes, and the sturdier, purple-blue clustered bellflower (6–10). Saw-wort (7–9) is one of the rarities. All in all it makes a poignant comparison with the surrounding 'improved' modern agricultural land. Even the turf of the many horse training gallops roundabout carries few if any wildflowers.

Old chalk grassland also carries a notable butterfly life, and a large number of species have been recorded here, including several which are rapidly declining along with the remaining areas of untouched chalk grassland which are not protected as nature reserves. The key here is the plant on which the butterfly lays its eggs. The chalkhill blue, for example, will only lay on horseshoe vetch, and is not seen when this plant is absent. The small blue chooses kidney vetch and the brown argus chooses common rock-rose. The dark green fritillary with a fast flight is worth looking out for in July. It is often seen, although the hairy violet, the plant it normally lays its eggs near, has not so far been recorded here. The marsh fritillary butterfly, now scarce in many of its old haunts, has recently colonised the site, patronising the devils'-bit scabious (6–10).

Management
The invading scrub is being removed and the flat turf is mown twice yearly in Spring and late Summer. Volunteers mow the disc and bowl barrows when necessary.

SEVEN BARROWS

OS sheet 174; SU 329828

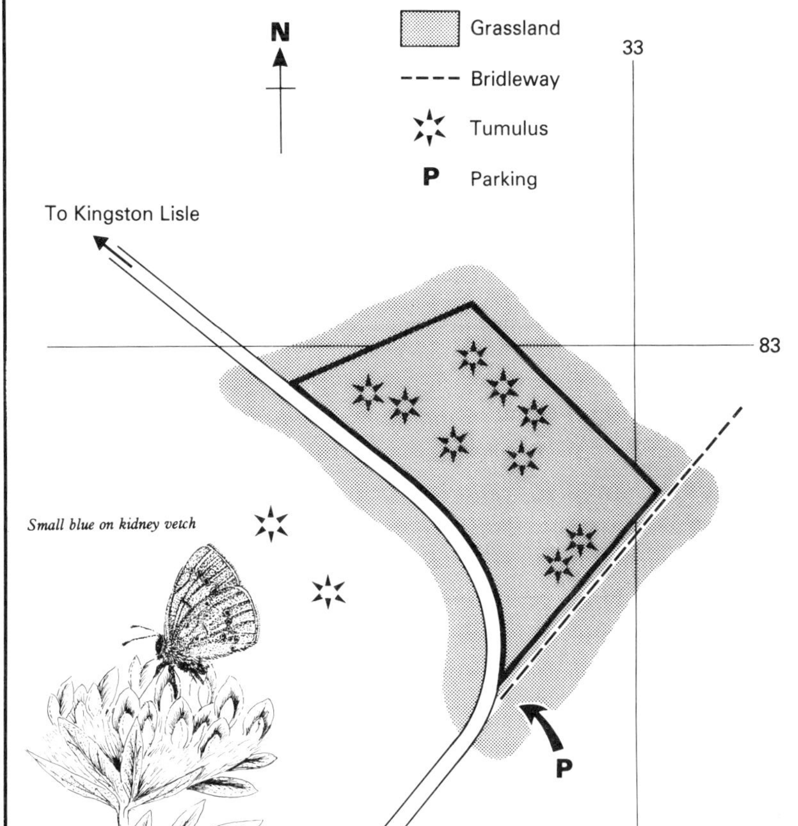

Small blue on kidney vetch

SOLE COMMON POND

OS sheet 174; SU 413707

Nearest town Newbury

A pond with adjacent areas of bog and heath and woodland, 4 hectares (10 acres) in extent, managed since 1975 by agreement with the owners.

Location
Leave Newbury to the west on the A4 and quite soon at the edge of Speen fork right along the B4000 to Wickham and Lambourn. After 2½ miles, turn right into the Boxford road and after about 400 yards you come to a track on the left. Park at the side. The reserve lies in the valley alongside the track as the plan shows. The reserve is unfenced; the boundaries follow the lip of the valley.

Access
Open to the public. Please avoid walking around the edges of the pond and up the floor of the valley on the sphagnum bog, which is quickly damaged. Dogs are not permitted on the reserve.

Description
The valley is fed by a small stream running from the sandy soil and provides just the right conditions of nutrient-poor, acid waterlogging for a bog to develop. Here we have one of the few remaining bogs in our three counties. It has been damaged by over-collection of the bog moss in the past, but is recovering under BBONT's management.

In the area of bog, round-leaved sundew can be seen (flowering 6–8), growing on its surface, and bogbean (5–7) and marsh St. John's-wort (6–9). Look also for the polypody fern.

The small pond itself is well worth studying – it is rich in wildlife within its border of mosses. What will be seen will reflect the time of year, but in the Summer months dragonflies are usually abundant.

Other habitats in the reserve are heathland on the northern slopes (being invaded by birch and bracken) while the southern slopes carry mature woodland with oak, birch, beech, sweet chestnut and some conifers.

Of the birdlife, the woodcock are perhaps most interesting for they are not all that common within BBONT's boundaries. They are waders which have deserted the open marshes to live in woodland. However they need wettish woods, for their long and rather sensitive beaks can only probe soft ground for worms and other food. When disturbed this bird twists away amongst the trees. Wood warblers are also heard here in Spring and Summer; notes accelerating and blurring into a shimmering trill every ten seconds or so, heard from the canopy above. They like woods such as this, with a rather open floor.

Management
Recent management has cleared large areas of the bog and heath of birch and a weir has been constructed to maintain the water at a constant level. Future plans include control of the birch and reducing the bracken on the heathland.

SOLE COMMON POND

OS sheet 174; SU 413707

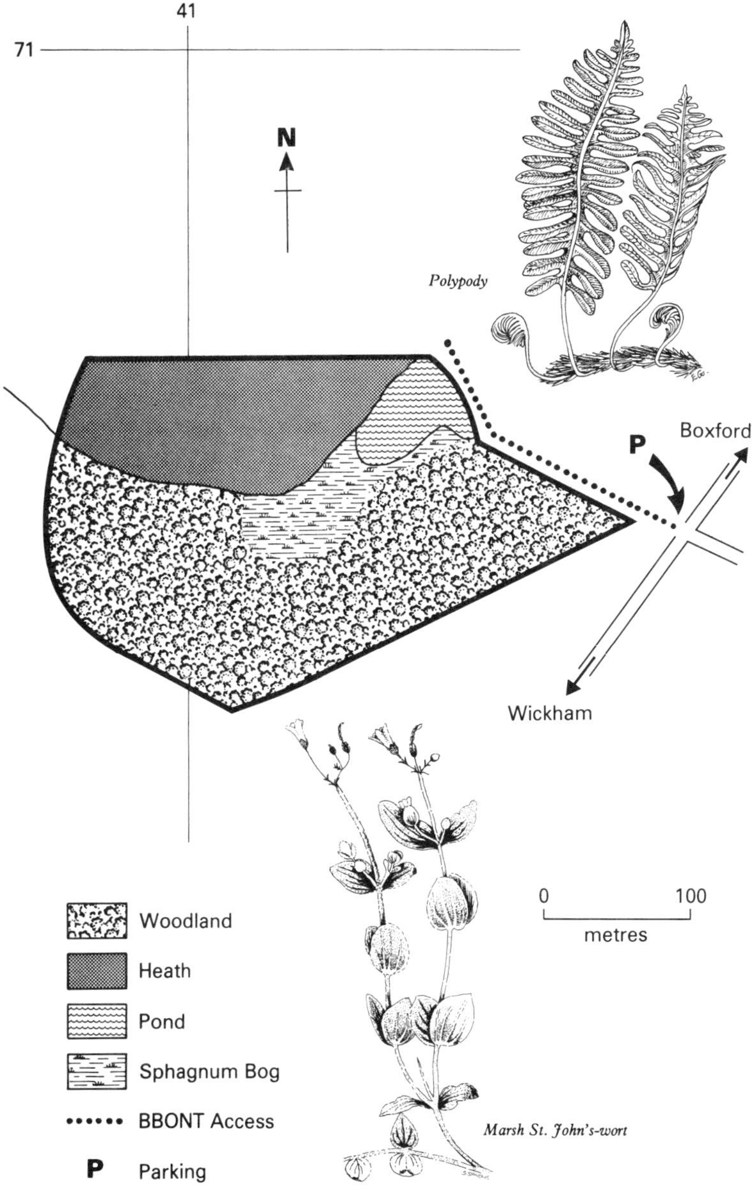

BUCKINGHAMSHIRE Reserves

Aston Clinton Ragpits
Bernwood Meadows
Boarstall Duck Decoy
Buckingham Canal
Burrows Reserve, Bernwood
Buttler's Hangings
Chequers
College Lake Wildlife Centre
Cowcroft
The Crong Meadow
Dancersend
Deep Mill Lane Pond
Gomm Valley
Grangelands and Pulpit Hill
Little Linford Wood
Long Grove Wood
Long Herdon Meadow
Millfield Wood
Oakley Parish Hedge
Oxford Lane
Park Wood
Pilch Field
Pitstone Hill
Rushbeds Wood and Lapland Farm
Stony Stratford
The Straits, Salcey Forest
Weston Turville Reservoir
Windsor Hill

ASTON CLINTON RAGPITS

OS sheet 165; SP 888108

Nearest town Tring

A small area of old chalk pits 2.4 hectares (6 acres) in total. An SSSI, BBONT owns 2.1 acres and manages the remainder under an agreement with Aston Clinton Parish Council.

Location
From Wendover take the A4011 towards Tring. After 2¼ miles, turn right onto an unclassified road towards St Leonard's. After 70 yards, park on the left hand verge. Entry to the reserve is by the kissing gate next to the double iron gates.

Access
Open to the public, but no dogs please.

Description
This is a small but varied area of old pits and spoil heaps at the foot of the Chiltern escarpment, once worked for superior quality chalk freestone or 'rag' used for local building. When work ceased, plants and animals from the countryside around colonised the site which now holds a rich variety of wildflowers and shrubs and animal life. Paradoxically, many of these species have now been driven from that wider countryside by modern intensive farming.

The woodland fringe to the west serves as a most useful screen. Beech and yew predominate. The rest of the reserve is grassland onto which scrub has invaded, mainly hawthorn (flowering 5–6) and dogwood (5–7) but there is also buckthorn (5–6), spindle (5–6), guelder-rose (6–7) and wayfaring-tree (5–6). Apart from their blossom, these shrubs are also handsome with berries later in the year.

Thirty species of butterfly have been recorded, many of them linked with the grassland which is a botanist's delight, a natural garden. The wildflower species include cowslips (flowering 4–5), bright blue chalk milkwort (5–6), purple autumn gentian (8–9), yellow horseshoe vetch (5–7), and nine species of orchids. These last are at their best in June and July.

There is a population of slow-worms, which have an appetite for slugs. Although there is a tale of a slug running away (in a manner of speaking), a slow-worm can move quickly enough when disturbed. It is not a snake but a legless lizard; sometimes a slight 'hip' can be noticed. It may shed its tail in an emergency in the way of lizards. A true snake, the harmless grass snake, may also be seen on the reserve.

Management
The aim is to maintain the variety of habitats here. The grassland within the pits has been fenced and is periodically grazed by sheep. Scrub has been cleared (the dogwood is a problem as it suckers when cut); mowing is carried out to stop scrub invading the open ground. Some lengths of hedgerow have been cut and 'laid'.

ASTON CLINTON RAGPITS

OS sheet 165; SP 888108

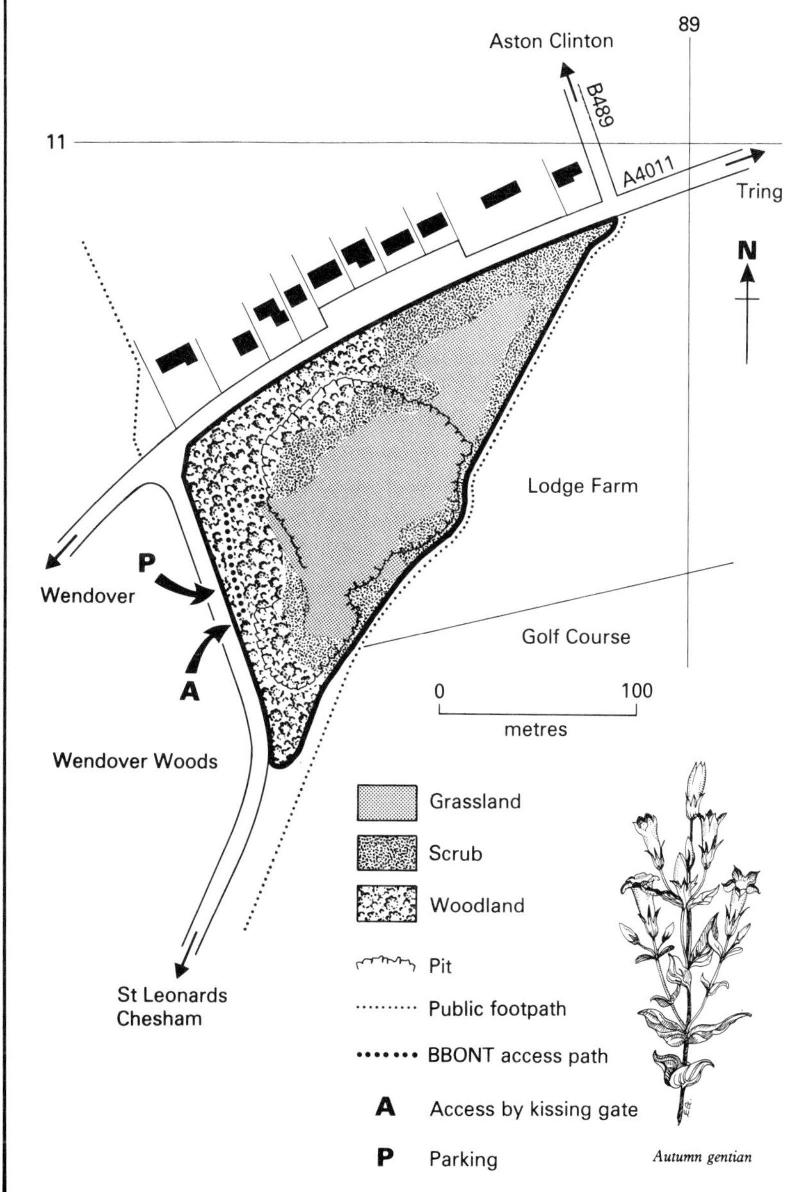

Autumn gentian

BERNWOOD MEADOWS

OS sheets 164 & 165; SP 606110

Nearest towns Oxford and Thame

Two flowery meadows bordered by hedges, some of which are ancient. There are also two small ponds. The reserve totals 7.3 hectares (18.3 acres). An SSSI, it was bought by BBONT in 1981.

Location
From Headington roundabout on Oxford's eastern ring road, take the road north past the crematorium; turn right, cross to the B4027 and proceed to Stanton St John. Take the first left in the village. Fork left at the road junction at Menmarsh, and after ¾ mile there is a small car park on the right.

Access
Open to the public but please avoid damaging the flowers and disturbing the butterflies by keeping to the two footpaths from mid-April until the hay has been cut. Dogs are not allowed when stock is grazing.

Description
Although medieval 'ridge and furrow' is to be seen, especially in the western meadow, the fields have seemingly escaped the plough in more recent times. Neither has the use of artificial fertilisers encouraged coarser growing grasses at the expense of other plants (certainly for the past 25 years at least the only manuring has been from grazing livestock). As a result of this and the hay and grazing regime, over 100 plant species have been recorded here, including 23 different wild grasses. There is a good display of green-winged orchid (flowering 5–6). Adder's-tongue can be seen – this fern is a useful 'indicator' plant of ancient meadows.

With the ground undisturbed for centuries, the communities of plants reflect differences in the soil below. The wide stripes of the medieval ridge and furrow were created by the ox teams ploughing down one side, turning at the 'headland' and ploughing back up the other, throwing the soil inwards on both passes to create the ridge. In early Summer buttercups colour the drier ridges but are absent from the furrows. By contrast, the furrows with their damper soil promote scattered growths of rushes and other damp-loving plants which are absent from the ridges.

The patch of lush green grass with few flowers, almost due south of the car park is therefore of some interest. The ground has clearly been disturbed and fertilised here and it is thought that it may be the grave of cattle slaughtered in a serious foot and mouth epidemic in the early 1960's.

Butterflies and other insects add further interest to these attractive meadows.

Management
The hay is mown in July, followed by grazing of the 'aftermath'. The hedges are being managed by coppicing on a long rotation, being cut at intervals of many years to encourage the regeneration of young blackthorn. The willows around the larger pond are pollarded.

BERNWOOD MEADOWS

OS sheets 164 & 165; SP 606110

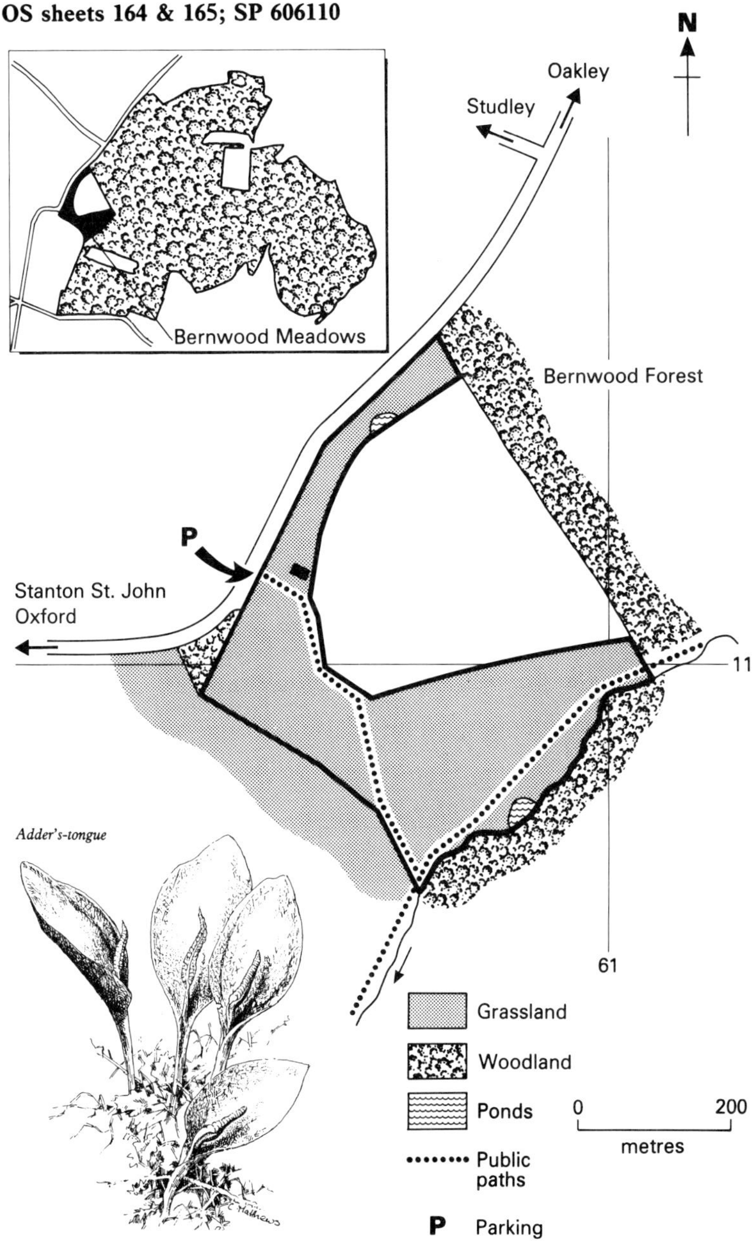

Adder's-tongue

BOARSTALL DUCK DECOY

OS sheets 164 & 165; SP 623151

Nearest towns Bicester and Thame

An historic man-made decoy pond surrounded by woodland, 6.6 hectares (16½ acres) in total. It is owned by the National Trust and managed by BBONT, on lease from the National Trust since 1980.

Location
Take the B4011 out of Thame through Oakley and then first left. After 300 yards turn right through Manor Farm and on to the Decoy Cottage where there is a car park. Take care through the farm as there are children; speed limit 10 mph.

Access
The reserve is open at weekends and bank holidays (10am–5pm) and Wednesday afternoons (2–5pm) from Good Friday to the August bank holiday inclusive. Admission is free to members of the National Trust and BBONT (you will have to show your card); there is a charge for non-members of the two. There is a small information centre and a nature trail. A guide book and checklist of birds seen at the reserve are on sale.

Description
This is one of three working decoys left in the country. A Dutch device to catch wild duck for the table, they were once not uncommon on country estates.

The shallow 2½ acre 'pond' has curving narrowing channels or 'pipes' running from it. Three of these are in good repair, and the home pipe is in use. This is covered by netting stretched on hoops and has, along one side of the channel, a row of overlapping reed screens tall enough to hide a man but with small gaps strategically left between them. Each gap is just large enough to allow a specially trained dog to squeeze through. In full view of the birds, the dog passes in front of the screen and then dodges out of sight again at the next gap. Rather than being frightened, the curiosity of the hapless waterfowl seems aroused for they swim closer. Another appearance further up the pipe lures them on until at last, in the trap net at the end they meet their fate.

Today however it is legs that are ringed, not necks wrung! During the Winter months many waterfowl are caught, ringed and released at Boarstall. One ring from here was returned from Turkey. Another was sent back by a restaurant in Soho. "Your pheasant (sic) was delicious" they wrote.

Decoy demonstrations are held at 11am and 3pm at weekends and bank holiday Mondays, subject to the availability of the warden.

There is a nature trail through the woodland around the decoy pond which can, with help, be used by disabled visitors with wheelchairs. This woodland is rich in wildlife – from March to July 1988, for example, 45 bird species were recorded of which 30 were probably nesting.

Management
This is carried out jointly by the National Trust and BBONT. A hide is planned. Some parts of the wood have been coppiced.

BOARSTALL DUCK DECOY

OS sheets 164 & 165; SP 623151

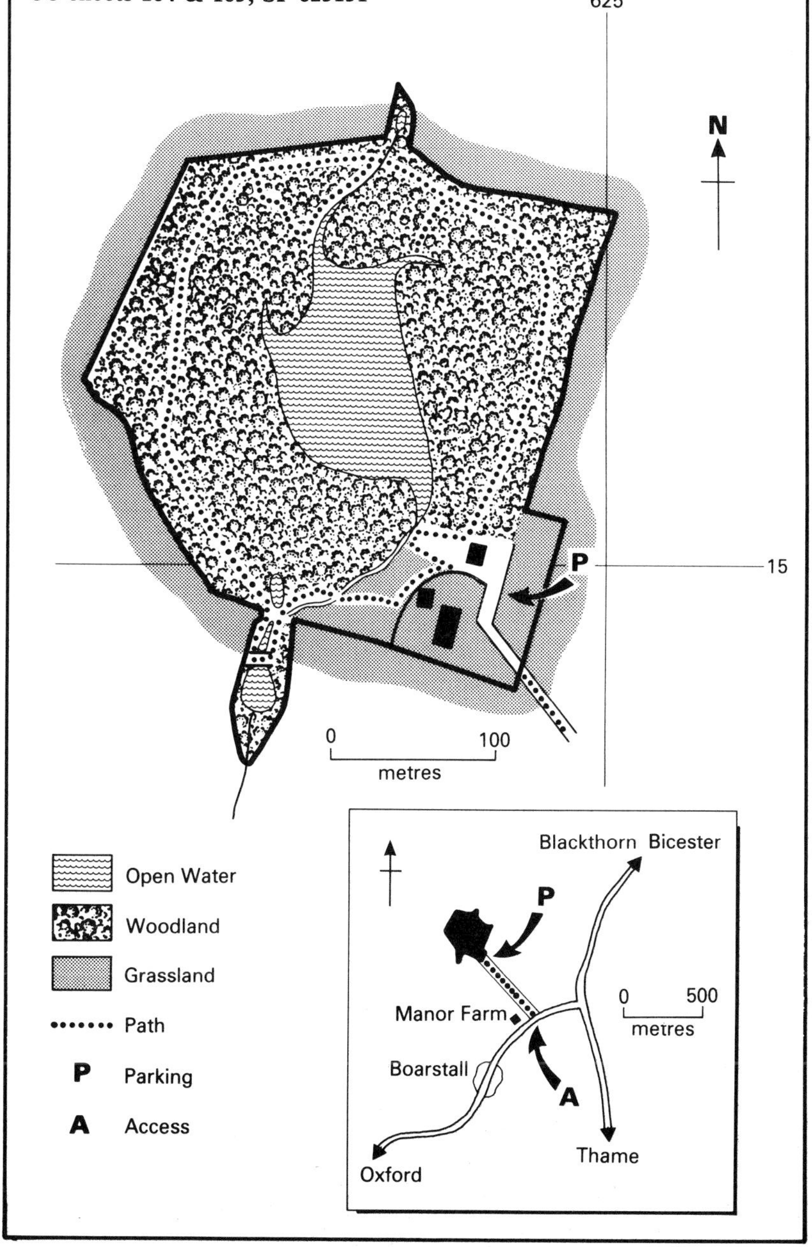

BUCKINGHAM CANAL

OS sheet 152; SP 725352

Nearest town Buckingham

A length of disused canal, 1 hectare (2.65 acres) in all. It was bought by BBONT in 1972.

Location
To reach the reserve, take the A422 east from Buckingham for 2 miles until you reach a farm track on the right, leading to a pumping house (Anglian Water). Members of the public should park near here and proceed on foot; BBONT members however may drive down the track, and just before the pumping house go through the gate on the left and park in the field adjoining Hyde Lane Gravel Pit. (If parking, indicate that you are a BBONT member). Then walk down to the pumping house and cross over to the canal towpath. The reserve begins in about 100 yards.

Access
Do not disturb the anglers at Hyde Lane Lake. Note that the sides of the lock are high and could be a hazard to unsupervised children.

Description
This reserve embraces both banks of the disused canal, from the point of entry along to the disused lock and on past it to a bridge.

'Waste' or now-disused sites are often of high wildlife interest, and this disused canal is a case in point. The canal bed at the western end is dry but pools of water are found towards the eastern end. Just beyond the lock, the overflow from the gravel pit creates a piece of open water. Beyond this the site becomes overgrown with emergent aquatic plants until, just before the bridge, there is again open water.

There are varied conditions here, shallow pools and deeper water, which provide a sanctuary for aquatic life. We must remember that shallow ponds are now hard to find in today's manicured countryside. Most of the old field pools have been filled in (if cattle or sheep are grazed, water is piped in). Many wetlands have been drained. In many areas, even the 'common' frog which prefers shallow water has now its breeding stronghold in the ponds of suburban gardens rather than in the countryside.

So here, we have a haphazard, sheltered aquatic habitat of a kind once common everywhere. Frogs, toads, and colourful damselflies and dragonflies make use of it in preference to the open water of the flooded gravel pit nearby. Grass snakes are also seen here – they swim well and hunt frogs and other aquatic life. Among the unusual birds recorded have been the bittern and water rail.

Management
Areas of open water need to be maintained, and future work needs to concentrate on such tasks as the removal of scrub along the canal banks and pollarding the willows near the lock. The hedge along the southern boundary needs to be trimmed down to let more light onto the water. The creation of more pools along the bed of the canal would be beneficial.

BUCKINGHAM CANAL

OS sheet 152; SP 725352

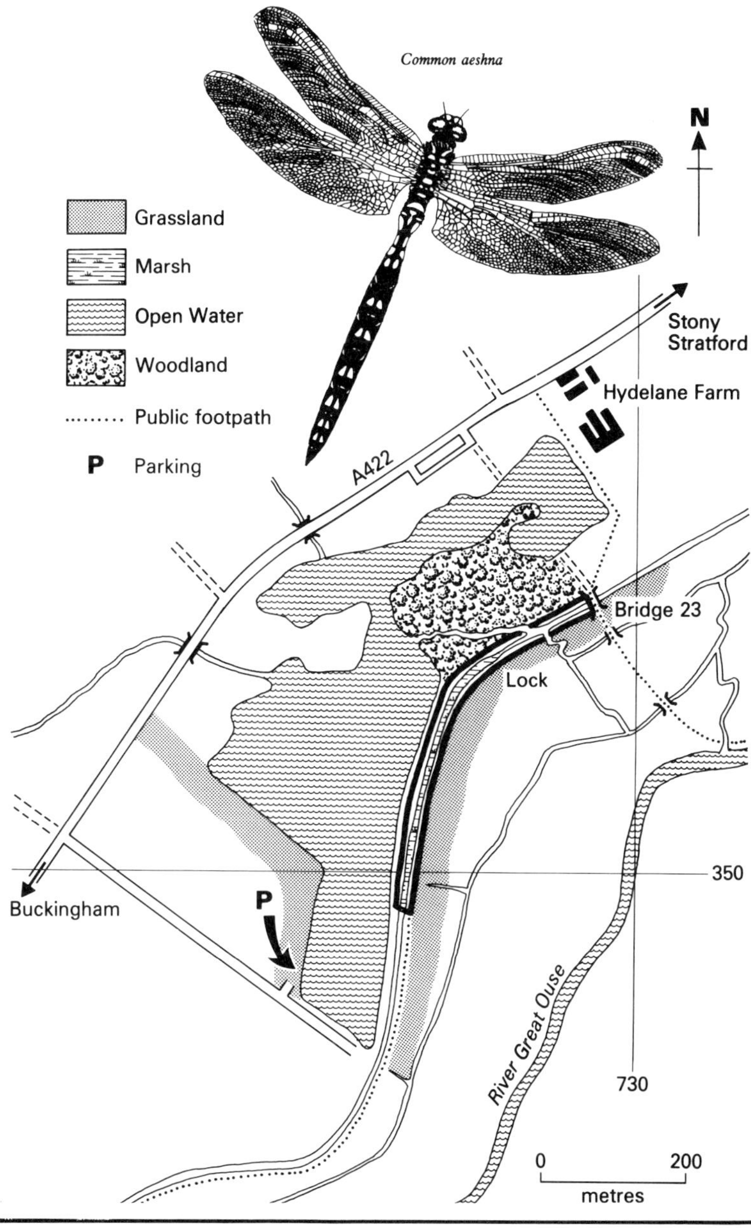

BURROWS RESERVE, BERNWOOD

OS sheets 164 & 165; SP 624114

Nearest towns Oxford and Thame

A small block of scrub with some oak trees, 0.29 hectares (0.725 acres) in extent. BBONT bought the land in 1966. It was notified as part of the Shabbington Wood complex SSSI in 1987.

Location
From Headington roundabout on Oxford's eastern ring road, take the road north past the crematorium; turn right, cross to the B4027 and proceed to Stanton St John. Take the first left in the village. Fork left at the road junction at Menmarsh, and after ¾ mile there is a small car park on the right. From here walk along the Forestry Commission ride in Oakley Wood for a mile and turn left along another ride. The reserve lies at the end.

Access
Open to the public.

Description
The reserve was bought for its colony of black hairstreak butterflies. Conservation must often be directed to provide for special needs of wildlife which are endangered, and this butterfly provides a good example.

The female lays her eggs singly on the end shoots of blackthorn (sloe) and the caterpillars eat the leaves. They spin up onto the side of a twig in late Spring (the year's weather influences the exact date) and change into chrysalids. The chrysalis closely resembles a bird dropping, an effective camouflage against predators in the shape of insect-eating birds. The adults are on the wing late in June and throughout most of July.

They are not very active butterflies, however, and are seldom noticed by accident; rather they have to be searched for. (The reserve is in fact named for the person who first discovered these butterflies in Bernwood).

Blackthorn is of course common enough, out of woodland as well as in it, but the butterfly is confined to the area lying generally between Oxford and Peterborough. It seems to prefer blackthorn growing thickly in sheltered but sunny places such as open woodland rides and clearings, and there are at least 3 BBONT grassland reserves with colonies in hedgerows.

Its rarity may in part be explained by its reluctance to fly far, to colonise new sites, and so guarantee its survival in the event of its home thicket becoming unsuitable in some way. In the much more varied countryside of the past, blackthorn thickets of suitable age were quite likely to be within easy reach.

Management
To provide blackthorn of varied ages, some of the overmature growth was cut and removed so that young healthy suckering shoots can grow up. Some of the oak and ash trees have also been felled because of the shade they cast on the young growth. Small fenced enclosures have been erected to prevent fallow and muntjac deer from browsing the young blackthorn.

BURROWS RESERVE, BERNWOOD

OS sheets 164 & 165; SP 624114

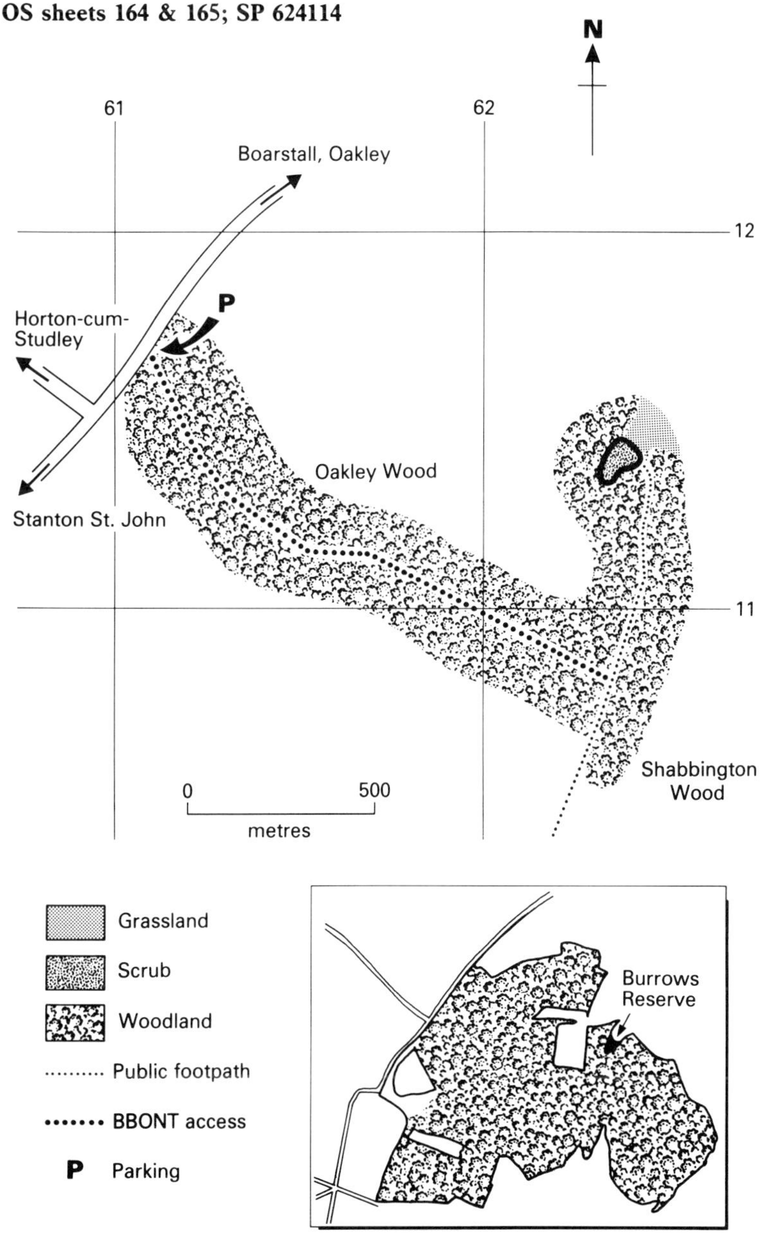

BUTTLER'S HANGINGS

OS sheet 165; SU 818962

Nearest towns High Wycombe and Princes Risborough

Two paddocks of chalk grasslands, 4 hectares (10 acres) in extent, with tongues of beech woodland extending into the reserve from the woodland above. It is an SSSI, leased by BBONT from the Dashwood Estate since 1974.

Location
Leave the A40 and West Wycombe via Chorley Road northwards to Bledlow Ridge and Chinnor. Where the road divides into three, park on the road verge and walk 300 yards up the right hand fork to where a public footpath crosses the road. The footpath going east leads to and crosses the reserve on the south-west facing slope of the valley and continues on into Hearnton Wood and Bradenham.

Access
Open to BBONT members only, except on the footpath.

Description
This chalk grassland reserve was ungrazed except by rabbits for over 20 years prior to BBONT's lease. Their grazing must have been light for the paddocks were invaded by a scrub of hawthorn, dogwood and other shrubs on the lower slopes. There is a good variety of wildflowers including pyramidal orchid which can be seen in early Summer (it flowers 6–8) and the purplish flowered Chiltern gentian is seen later (8–9). A Chiltern speciality, wild candytuft, flowers (7–8) at the top of the slope.

The reserve has interesting invertebrate life. 25 species of butterfly are regularly on the wing in the two paddocks, including the chalkhill blue (which is linked to horseshoe vetch, often found in old grassland of this kind), the marbled white (the caterpillars of which are grass feeders) and the dark green fritillary. The close relatives of the last are seen in woodland, but the dark green fritillary has adopted more open spaces, not only downland but also moorland. It is interesting that it still lays its eggs on (and relies on the presence of) wild violets, as do the woodland fritillaries.

108 different species of spider have been recorded here, including 2 national rarities. The great green bush-cricket is also present. Bush-crickets are related to grasshoppers, but with long antennae or 'whiskers' (they were often called 'long horned grasshoppers'). The great green bush-cricket has a penetrating song, a feature of warm Summer nights on the reserve.

Management
BBONT volunteers carried out scrub clearance in the early years and current management aims to restrict any further invasion of scrub and leave the grassland open to the sunshine. Winter grazing by the BBONT sheep flock is planned for both paddocks. This will improve the sward for the benefit of chalk grassland plants and invertebrates.

BUTTLER'S HANGINGS

OS sheet 165; SU 818962

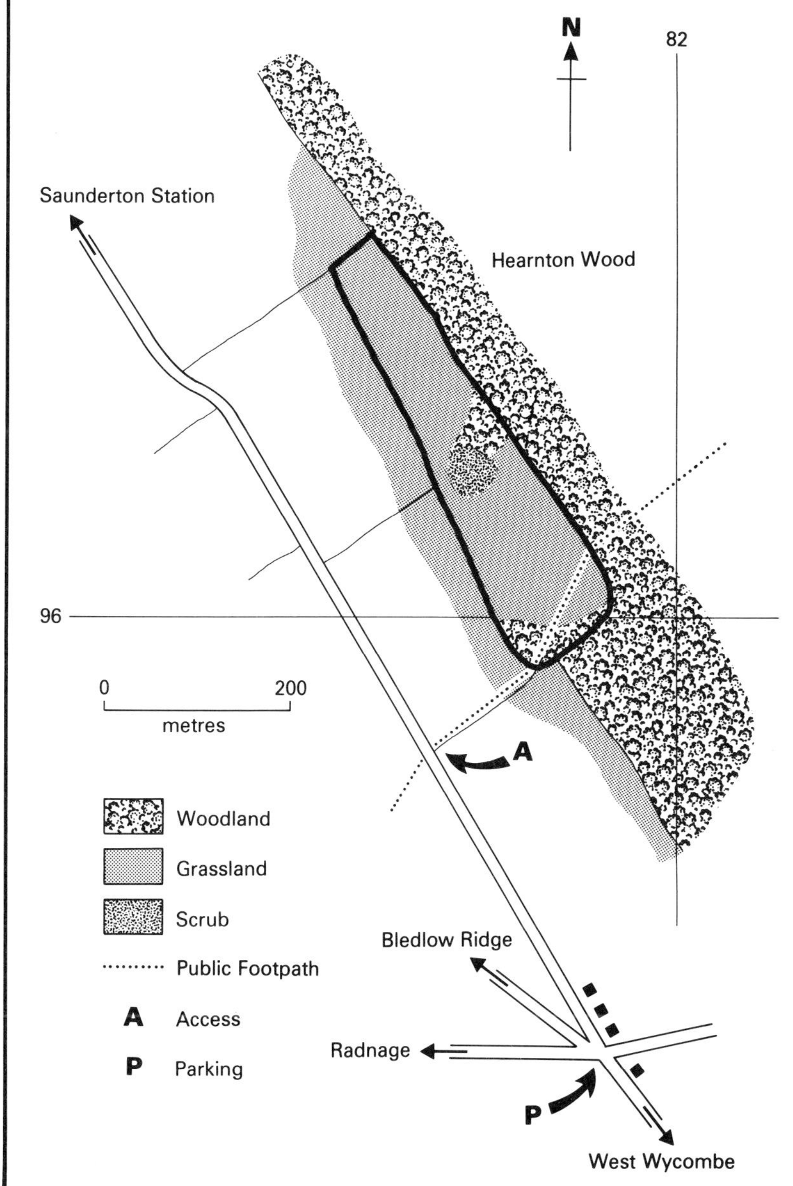

CHEQUERS

OS sheet 165; SP 830055

Nearest towns Princes Risborough, Wendover and Aylesbury

Very rich chalk grassland on the steep flanks of the Chilterns, totalling 33.2 hectares (83 acres). It is part of the Ellesborough and Kimble Warren SSSI and has been managed by BBONT by agreement with the Trustees of the Chequers Estate since 1972.

Location

From Princes Risborough take the A4010 north to Great Kimble and park in the lay-by on the right hand side just before the Bernard Arms. Walk up the bridleway to the right for ½ mile and enter the reserve along the footpath to the left. An alternative entry to the site is from the adjacent GRANGELANDS reserve (see page 86).

Access

Open to the public, but a permit, obtainable from the BBONT Office in Oxford, is required for visits to the coombe area.

Description

The grassland matches that of other BBONT reserves on chalk described in this handbook. It carries a galaxy of chalkland plants, especially along the eastern end of the coombe. Some to look for are common rock-rose (flowering 5–9), wild thyme (5–8), the rare musk orchid (6–7) and squinancywort (6–7). The odd name of this last plant has a convoluted origin. 'Wort' is an old word for any soft green plant. 'Squinancy' is a muddled form of 'quinsy' – the old name for tonsillitis. An infusion of the leaves of this plant was thought to cure sore throats if gargled. In olden days when most children died of diseases such as tonsillitis before they reached the age of 15, such remedies had almost religious sanctity. There is however no proof that this cure (or many others) was effective.

Box (3–5) grows in the coombe and also on Ellesborough Warren at the north-east of the SSSI. This is thought to be one of the only three truly native box sites in Britain; it has been planted elsewhere however. Deadly nightshade (6–8) also grows on the reserve and it is worth coming to see. Its English name is familiar to all and its scientific name is *Atropa belladonna* – 'beautiful woman' in Italian. You can speculate why Victorian botanists awarded it this accolade. "Banish it from your gardens", wrote a 16th century herbalist: "being a plant so furious and deadly". It is in fact a cousin of the potato, but a handful of its shiny cherry-like berries can kill. The juices affect the nervous system, weakening the pulse, racing the heart.

The history of plant names can be of some interest. There is also the mound of an ancient fort to be seen here, and a sunken way. So here, while botanising historically you can also walk the steps of time in another way.

Management

The grassland is maintained by grazing. Ragwort, poisonous to livestock, is removed by volunteer work parties.

CHEQUERS

OS sheet 165; SP 830055

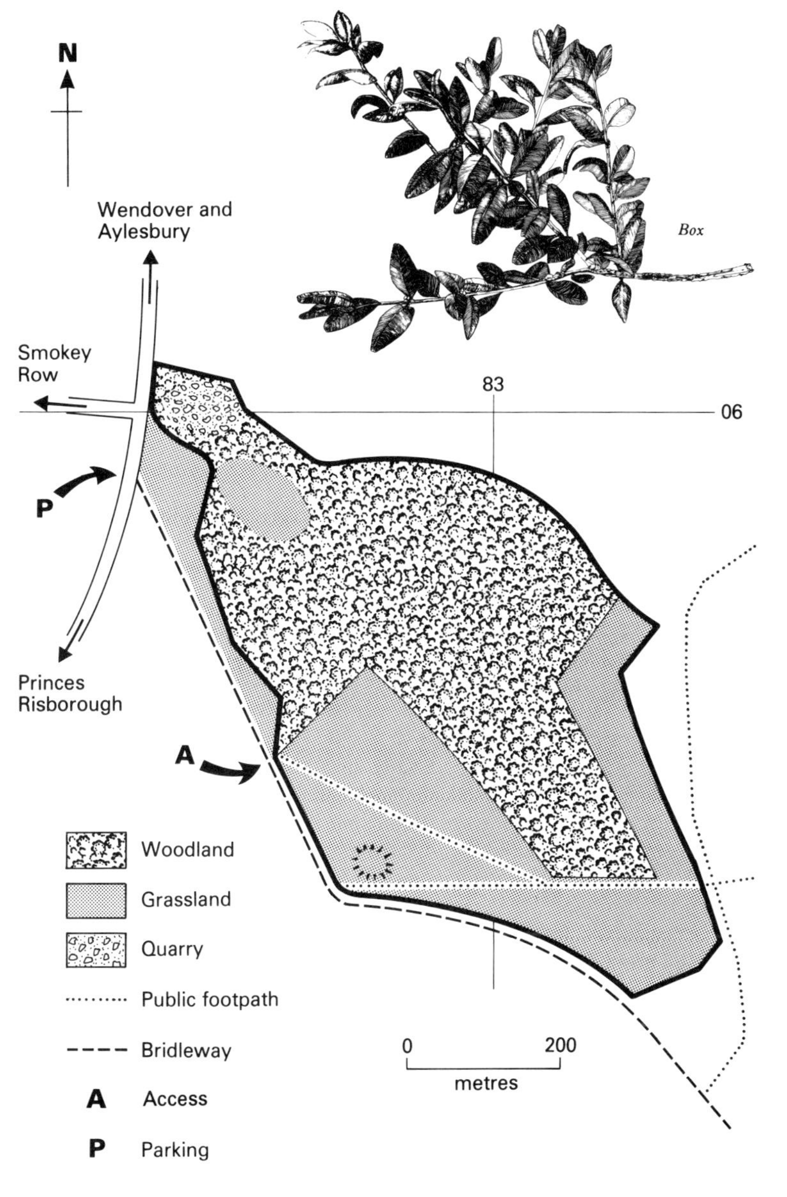

COLLEGE LAKE WILDLIFE CENTRE

OS sheet 165; SP 935139

Nearest town Tring

A worked-out chalk quarry which is being restored as a wildlife centre, a joint project between BBONT and Castle Cement (Pitstone) Ltd. 40 hectares (100 acres) in area, it includes a 25 acre lake. Part of the quarry has been designated an SSSI for its geological interest.

Location
From the roundabout on the A41 just to the west of Tring, take the B488 north-eastwards past Tring to 200 yards past the canal bridge at Bulbourne. The entrance is on the left hand side.

Access
All visitors must have a permit. The site is open to the general public on Sundays, 10am to 5pm, with permits from the Warden's Office on the reserve. BBONT members can visit at any time with a monthly or annual permit from the same office or from the Conservation Officer, Castle Cement (Pitstone) Ltd, Pitstone, Leighton Buzzard, Beds LU7 9AR (please include a stamped addressed envelope).

Description
This is an excellent example of creative conservation in action. The rehabilitation plans have been devised jointly by BBONT, supplying expertise, voluntary labour and wardens, and Castle Cement who provide funding and machinery.

The project began in 1985. It now incorporates a lake, large islands, shingle beaches, floating nesting rafts, bird observation hides and a visitor/information centre.

It is particularly interesting for its birdlife, and is an addition to the Tring complex of open waters which are famous amongst bird watchers.

A variety of waders, terns and other species are regularly seen on passage while breeding birds include lapwing, redshank and little ringed plover. This last was seldom seen in Britain before World War II, but in 1938 a pair nested at one of the Tring reservoirs and since then it has become established, with hundreds now nesting at gravel pits throughout central and eastern England. Its eggs are laid in a simple scrape on the gravel, and the chicks are well camouflaged; nevertheless predation by crows and other pirates is heavy.

Aålarge woodland and a marshy area are also planned together with the extensive planting of trees and shrubs generally.

A sheep breeding programme is being established to promote interest in and further the work of the Rare Breeds Survival Trust.

The 5 acres occupied by the Arable Weed Research Centre are of great interest. This is a working museum of a kind, cultivating traditional wheat and other cereals together with once common cornfield flowers.

Management
The quarry banks are grazed by BBONT's sponsored flock of sheep and other rare breeds owned by Castle Cement.

COLLEGE LAKE WILDLIFE CENTRE

OS sheet 165; SP 935139

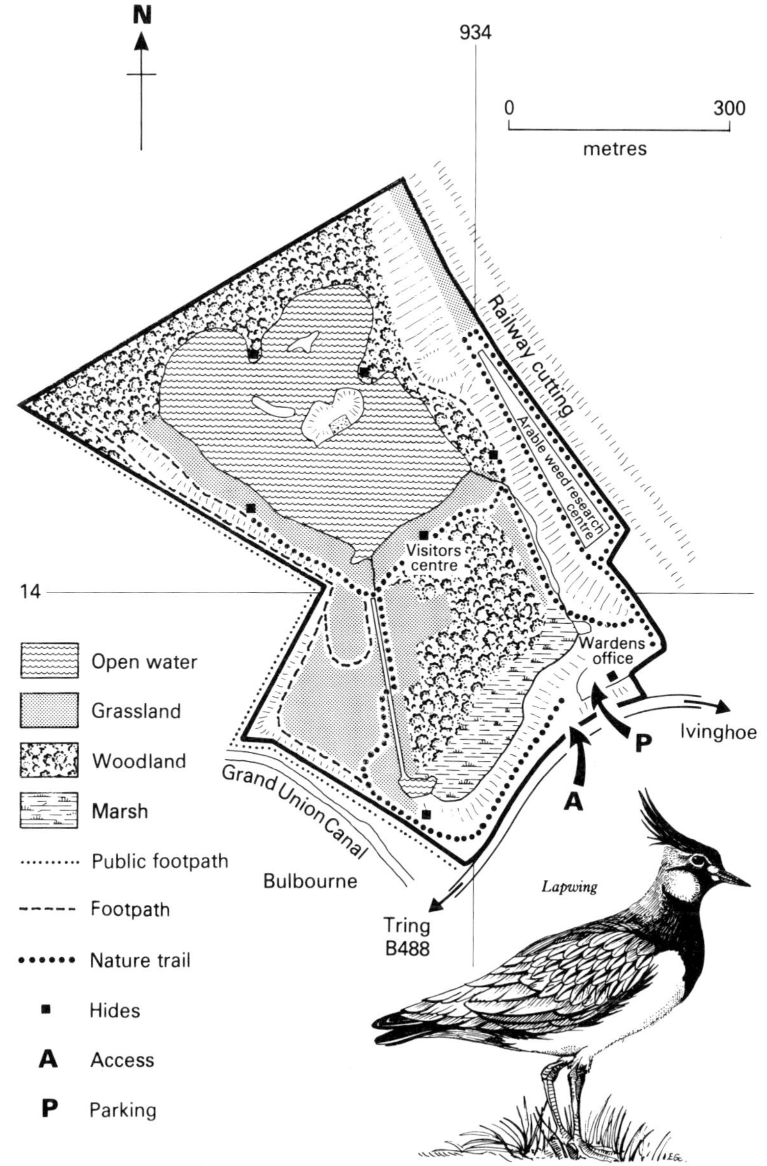

COWCROFT

OS sheet 165; SP 986018

Nearest town Chesham

An area of grassland and scrub, 0.6 hectares (1.5 acres) in area. The freehold was given to BBONT by Messrs Harman (Cowcroft) Ltd in 1980.

Location
Leave Chesham on the unclassified road through Botley to the T-junction at Ley Hill, turn right and park where indicated on the map. The reserve lies a quarter of a mile down the track, just before Cowcroft Wood.

Access
Open to the public.

Description
This area of grassland and scrub adjoins the old brick pit on its south side and Cowcroft Wood on its west side. The site is on disturbed calcareous (chalky) soil, and supports a great variety of flowering plants. Altogether, over 160 species have been recorded, including common centaury (flowering 6–10), eyebright (7–8), common spotted-orchid (6–8) and bee orchid (6–7).

An orchid site is always worth visiting. The strange beauty of these wildflowers has attracted generations of nature lovers and it is interesting to speculate on the reasons which may explain why they are so abundant in some places and years but not others, even when the soil and other conditions seem to be very similar.

Wild orchids are plants with a fragile and complex lifestyle. They produce large amounts of very small dust-like seeds which are widely spread by the wind, and so there is a good chance of them reaching suitable soil. The majority of our orchids prefer a warm dry climate (and are commoner in Europe than here). Many of them grow on southern chalk soils.

Being small, their seeds carry little food reserves and growth is very slow – it may be two or three years before the first small green leaf appears. In the meantime, the young orchid seems to rely on the presence of a helpful soil fungus which invades the roots of the young plant. This fungus continues its habit of obtaining nourishment from decaying matter in the soil, however, and the orchid takes its share. It may not be able to colonise new ground without the presence of the fungus.

The bee orchid which can be seen here has additional interest. Its velvet-textured lip resembles the back of a bumble bee, and the story has it that the drones, the male bumbles, are misled into thinking this a female, try to engage it, and so pollinate the flower. However it is also said that the bumble bee in question is a European, not British species! It is interesting to see what insects, if any, are visiting the orchids.

Management
This consts mainly of keeping the scrub under control.

COWCROFT

OS sheet 165; SP 986018

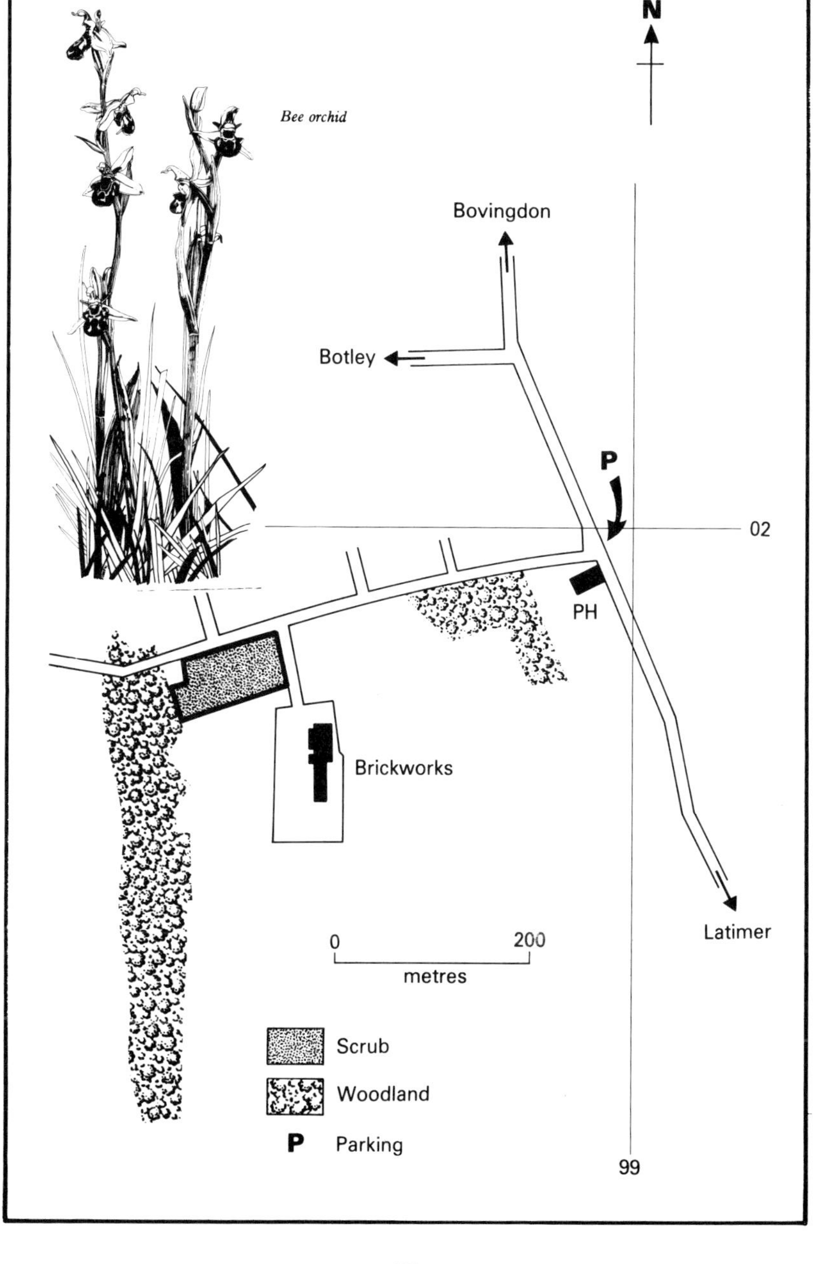

THE CRONG MEADOW

OS sheet 165; SP 904088

Nearest towns Wendover and Tring

An area of chalk grassland, about 1 hectare (2.6 acres) in extent. It has been held since 1987 on a 50 year lease from Thames Water.

Location
Leave Wendover on the A4011 towards Tring and turn sharp right just before the A41, onto the unclassified road leading to St Leonards. After 2½ miles you will find the waterworks buildings on your left. You may park in the yard at the back (please make sure that you do not block the gates or garages). The reserve lies on the opposite side of the road. (The reserve is adjacent to DANCERSEND reserve, see overleaf).

Access
Open to the public, but no dogs while stock are grazing.

Description
This is an example of 'unimproved' chalk grassland – ie grassland which has perhaps never been ploughed, or if so, not for many years. It has also escaped spraying with herbicides or feeding with artificial fertilisers. It was sold to Thames Water in the 1940's, since when it has been spasmodically grazed by horses or beef cattle.

The fact that it has not been ploughed or improved means that it contains a good variety of wildflowers and other plants – over 100 have so far been recorded. Many of the plants here are food and nectar sources for a range of butterflies including, among others, marbled white, ringlet and meadow brown. There are also reports of the white admiral butterfly. This is a woodland species with a liking for sunny glades. Presumably it comes from the woods of Dancersend reserve alongside, to find nectar at the flowers of the open ground.

Haphazard grazing in the past has resulted in the invasion of hawthorn and other scrub species out onto the open ground. Reserve management at a place such as this entails management of the grazing to allow the distinctive plants to grow. Undergrazing will invite scrub to invade the area. Overgrazing might control the scrub but could also wipe out colonies of flowers. Grazing at the wrong time of year (before seeding for example) might endanger some species of wildflowers in the long term (and it might also jeopardise the survival of the many butterflies and other invertebrates which live on such grasslands).

The livestock used for the grazing is also important as they feed in rather different ways. Sheep graze closely to produce a fine low sward. Cattle and horses tear at the grass. Cattle avoid dung-contaminated ground. Cowpat sites in their pastures will in time produce lush grass often signposted by a thistle as straight as a flagpole. If the heavier horses and cattle graze when the ground is soft after heavy rain, there is danger that it will be 'poached', churned to mud.

Management
A careful grazing regime is essential.

THE CRONG MEADOW

OS sheet 165; SP 904088

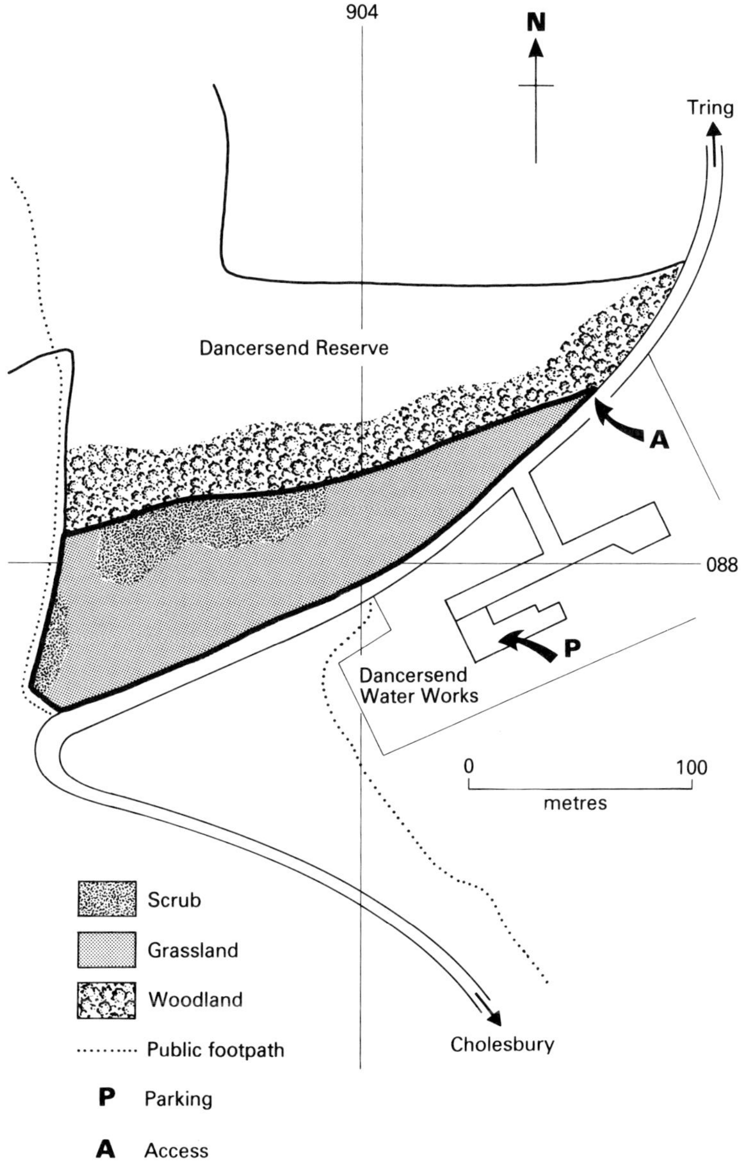

DANCERSEND

OS sheet 165; SP 900095

Nearest towns Wendover and Tring

30 hectares (79 acres) in all, made up of 73 acres of woodland and 6 acres of grassland and scrub, owned by the Royal Society for Nature Conservation and leased to BBONT. The whole reserve is part of a larger SSSI.

Location
From Wendover take the A4011 towards Tring and turn sharp right just before the A41 onto the unclassified road to St Leonards. After 1½ miles, at a sharp left hand bend, park on the right verge or in the pull-in on the left and walk up the track for 300 yards to the reserve.

Access
Open to the public along footpaths and trails, as shown on the plan.

Description
The reserve includes a range of Chiltern habitats on the slopes of two dry valleys. The soil varies, being quite acid in the higher woodland and alkaline in the lower grassland areas. Differences in soil, aspect and management have given us a rich variety of wildlife, with rare moths and butterflies and over 270 different plants.

Most of the woodlands were beech with some oak, but were felled in World War II, and left to be colonised by scrub. In the mid 1950's, most of the area was replanted by the Forestry Commission with beech, with larch and spruce as nurses. The rest was allowed to grow with self-sown trees, mainly ash. The conifers are now gradually being removed to reveal a remarkable number of survivors from the original woodland, such as wood vetch (flowering 6–8), stinking hellebore (1–4), and solomon's seal (5–6). There are also interesting fungi in Autumn, including the earth-star. Such a large area of woodland provides cover for a good many animals, including foxes, muntjac and roe deer and the fat or edible dormouse, introduced as a novelty into Tring Park in 1902, and now found in various parts of the Chilterns.

The central 12 acres, shown as open grassland on maps of over 100 years ago, have developed into a mixed woodland and scrub, very rich in plants, including the unusual yellow bird's-nest (6–8). New clearings encourage the grizzled skipper, Duke of Burgundy and other butterflies.

The jewel of the reserve is however a small area of chalk grassland saved from complete scrub invasion in the 1960's, when meadow plots receiving different management regimes were marked out. These now carry pyramidal, fragrant and bee orchids (at their best in June), clustered bellflower (7–8), Chiltern gentian (8–9) and other wildflowers and many butterflies including the green hairstreak and dark green fritillary.

Management
Blocks of woodland are being thinned or coppiced and clearings created. An area of scrub is being cut on a short rotation. The meadow plots are kept free of invading scrub and different heights of vegetation are obtained by a mixture of mowing and sheep grazing.

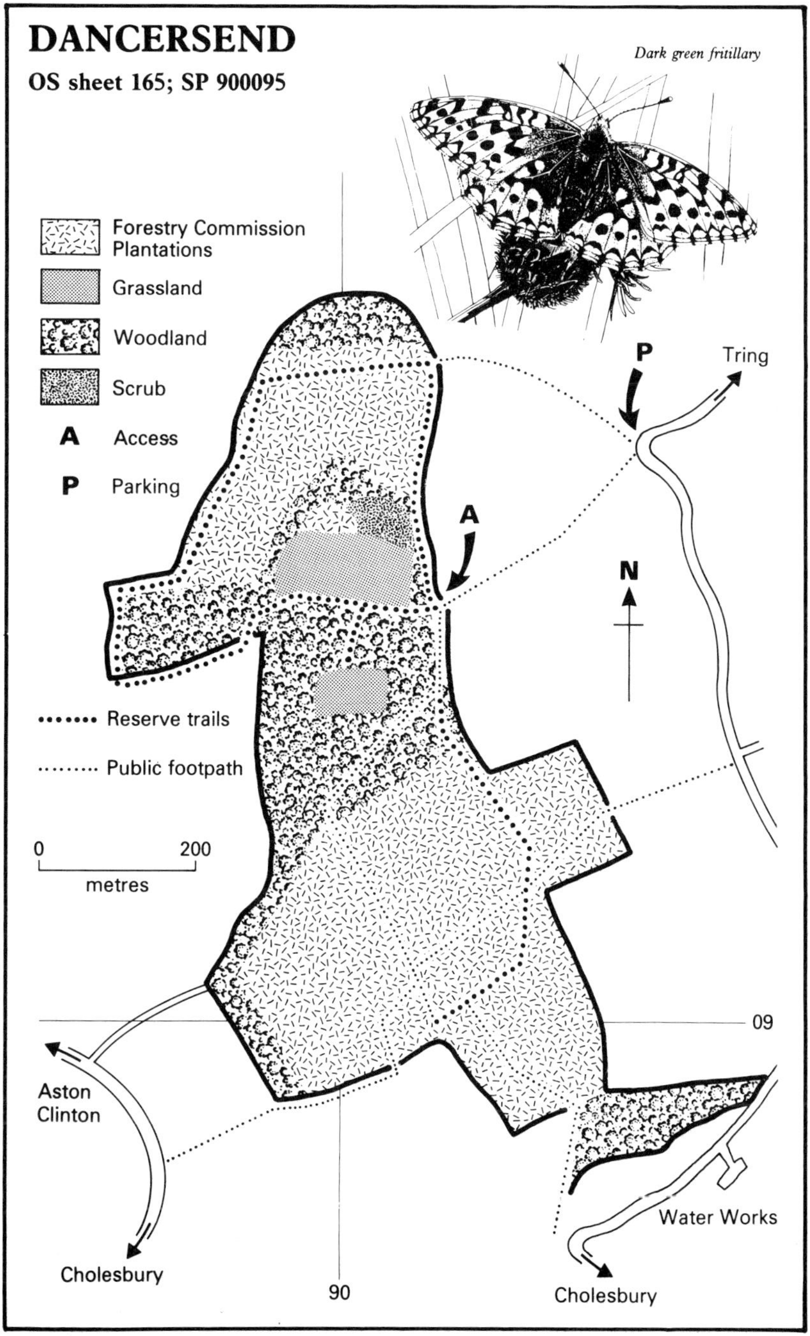
DANCERSEND
OS sheet 165; SP 900095
Dark green fritillary
Forestry Commission Plantations
Grassland
Woodland
Scrub
A Access
P Parking
P
Tring
A
N
Reserve trails
Public footpath
0
200
metres
09
Aston Clinton
Cholesbury
90
Water Works
Cholesbury

DEEP MILL LANE POND

OS sheet 165; SU 907995

Nearest towns Great Missenden, Amersham and High Wycombe

A rather varied reserve of 1.6 hectares (3.8 acres) comprising a pond fed by a small chalk stream, an adjoining spinney and a hazel hedge. BBONT has managed the site by agreement with Buckinghamshire County Council since 1983.

Location
From Wendover take the A413 past Great Missenden, turn right immediately after the railway bridge and park on the roadside at SU 907995.

Access
Open to the public, but there is limited parking. Please park with care and avoid blocking the entrance to the gas holder.

Description
The pond covers about 3½ acres, and is fed by the River Misbourne, which is here still a chalk stream. The pond may hold little water or even run dry at some times of year.

Although the pond is quite rich in aquatic wildlife, it has not yet been studied in detail. However, the first dip of an exploratory net produced one adult and four young smooth newts.

In spite of their poor press, newts are interesting wildlife. For much of the year they live on land on a diet of worms, snails and insects. They hibernate here and soon after awakening they move to ponds and assume nuptial dress. For the male smooth newt this is a high crest and an orange, darkly spotted belly. The female assumes a pale belly.

Newts mate at the bottom of the pond. The male approaches the female in lively fashion. Perhaps startled, she moves away and a chase ensues until, yielding, she stands still. The male moves across in front of her, displaying crest and colourful flank. He then faces her and lashes his tail, sometimes so strongly that the stream of water knocks her over. This fanning carries his smell to her and seems to trigger her mating instinct for she then approaches him. He turns and creeps away, but suddenly stops. She bumps into his tail, and he drops his sperm bundle and then turns to block her path, so that she ends with her nose against his tail. That seeming accident is important for it positions her exactly, to be able to take in his sperm bundle. And the explanation for all this? Well, their tails would make clasping each other in the manner of their frog cousins difficult.

Dragonfly and damselfly are also seen here. Their nuptials are as exotic as that of newts and end with the male grasping the female 'by the scruff of her neck'. In this 'tandem' position they mate. The eggs thus fertilised, the pair may stay in tandem while they are laid. Otherwise you may see the female skimming the surface, touching it lightly as she releases an egg. Other species perch to lay on aquatic plants.

Management
The roadside hedge would be improved if gaps were filled.

DEEP MILL LANE POND

OS sheet 165; SU 907995

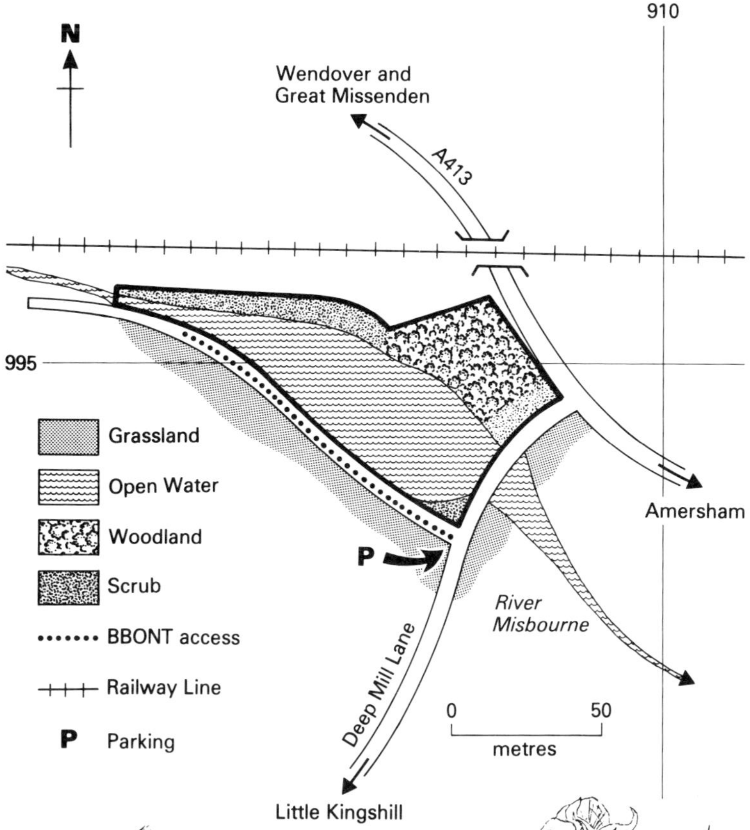

Yellow iris and reedbed

GOMM VALLEY

OS sheet 175; SU 898922

Nearest town High Wycombe

A tract of once open grassland invaded by scrub, 4 hectares (10 acres) in extent. An SSSI, it is managed by BBONT by agreement with the owners dated 1969.

Location
From the centre of High Wycombe take the A40 eastwards towards Beaconsfield for about 1½ miles, turn left up Gomm Road and park before or after the bridge. Take the public footpath diagonally across the slope to the first stile and then follow the lower path along the foot of the grassy slope and through the hedge into the next field. After 30 yards the path enters the reserve on the right.

Access
Open to the public.

Description
The history of the site has not yet been traced, but on the evidence of other dogwood scrub sites it was possibly once arable land.

It is in fact an excellent example of what is known as 'southern chalk scrub', being dominated by dogwood (flowering 5–7) and hawthorn (5–6) but with buckthorn (5–6), spindle (5–6) wayfaring-tree (5–6), so called because John Gerard, a 16th century botanist, wrote that it was the most common shrub the wayfarer would encounter, and guelder-rose (6–7) also to be seen. The shrubs provide a good display of blossom in season followed by a fine show of berries in the Autumn. These attract many birds, especially the thrush family, in Winter.

In the few small remaining grassy areas the common spotted- (flowering 6–8), bee (6–7) and pyramidal (6–8) orchids and common twayblade (6–7) grow.

In the small area of mixed broadleaved woodland at the southern end of the reserve grow bluebells (4–6), wood anemone (4–6) and the dainty coralroot with rose-pink flowers (4–5). In Summer, look for small brown bulbils at the junction of the upper leaves of this last plant. It is at the edge of its range in Britain and rarely sets seed, or if so, the seeds rarely ripen. As a strategy which increases the chance of survival, it produces these bulbils which drop to the ground to grow into new plants.

The reserve is noted for its butterflies, moths and plant galls – over 30 different species of butterfly, 180 of moth and 24 species of gall have been recorded here. The moths include the bee hawk-moth, a day-flying moth which mimics the bumble bee and so gains protection. In the long July dusks, glow worms may be seen. The wingless female beetle emits cold, far-reaching light which attracts the winged male.

Management
The scrub must be kept under control; dogwood regenerates vigorously if cut in the Winter and so Summer cutting is to be adopted as part of the future management of this reserve.

GOMM VALLEY

OS sheet 175; SU 898922

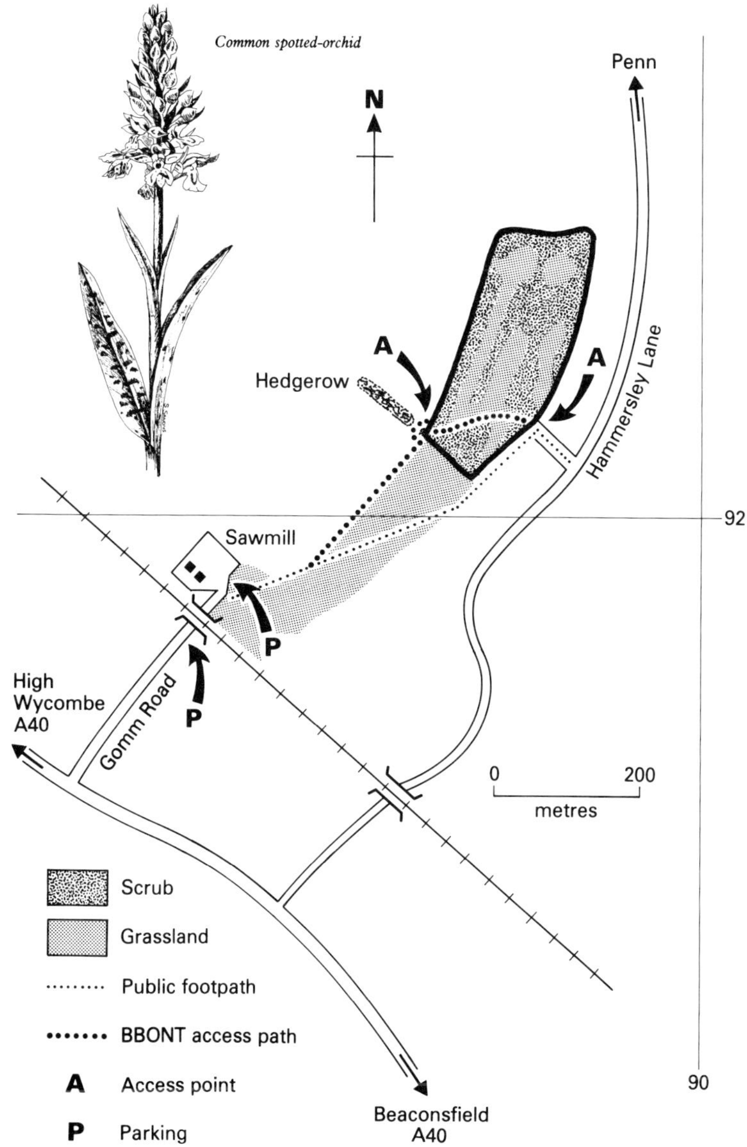

GRANGELANDS and PULPIT HILL

OS sheet 165; SP 827049

Nearest towns Princes Risborough, Wendover and Aylesbury

A 20 hectare (50 acre) area of chalk grassland with some scrub, and with beech woodland on Pulpit Hill. BBONT manages part of the Grangelands site by agreement with Buckinghamshire County Council; Pulpit Hill is now National Trust land, and BBONT manages this in conjunction with them.

Location
From Princes Risborough take the A4010 in a north-easterly direction for 1 mile and turn right towards Great Missenden. Follow the road up the hill for about 1 mile to the car park on the left hand side. The reserve is reached down the bridleway on the left hand side. [The CHEQUERS reserve can be reached by following the path beyond the disused rifle range.]

Access
Grangelands, including the disused rifle range, is open to the public; some of Pulpit Hill is very steep and impenetrable where thickly overgrown with scrub.

Description
Grangelands has a lot of interest. The ancient, flowery chalk sward was ploughed up during World War II, but then reverted to grass and it has since recovered remarkably well. It has been recolonised, so that there is now a good display of plants and butterflies to be seen. The juniper seems to be doing well. It is now a rare shrub in Britain, but here it is successfully regenerating.

The marbled white is one of the butterflies found here. On the wing in July and August, the female has the habit, unusual amongst butterflies, of dropping her eggs casually amongst the grass, making no attempt to attach them to the grass stems or leaf blades on which the caterpillars will feed. This may perhaps be noticed on a visit. Whereas other newly hatched and hungry caterpillars would be at a disadvantage, those of the marbled white eat their egg case and then retire soon afterwards to hibernate deep within a grass tussock, to wake to feed on the fresh growth on warm days after the turn of the year. The seemingly haphazard actions of wildlife – the things we might find puzzling on a walk – often have an explanation if we can learn the whole story.

On this reserve, the chalkhill blue butterfly has also been studied, as have the glow worm and various bumble bees.

Pulpit Hill is basically beech woodland with whitebeams and yew.

Management
Much of the grassland is kept as short turf by rabbit grazing and human trampling. Much scrub control is needed, especially at the bottom of Pulpit Hill. However, areas of scrub in the valley bottoms are being retained to provide shelter and nesting sites for birds.

On Pulpit Hill, some over-mature beech has been felled and replanted.

GRANGELANDS and PULPIT HILL

OS sheet 165; SP 827049

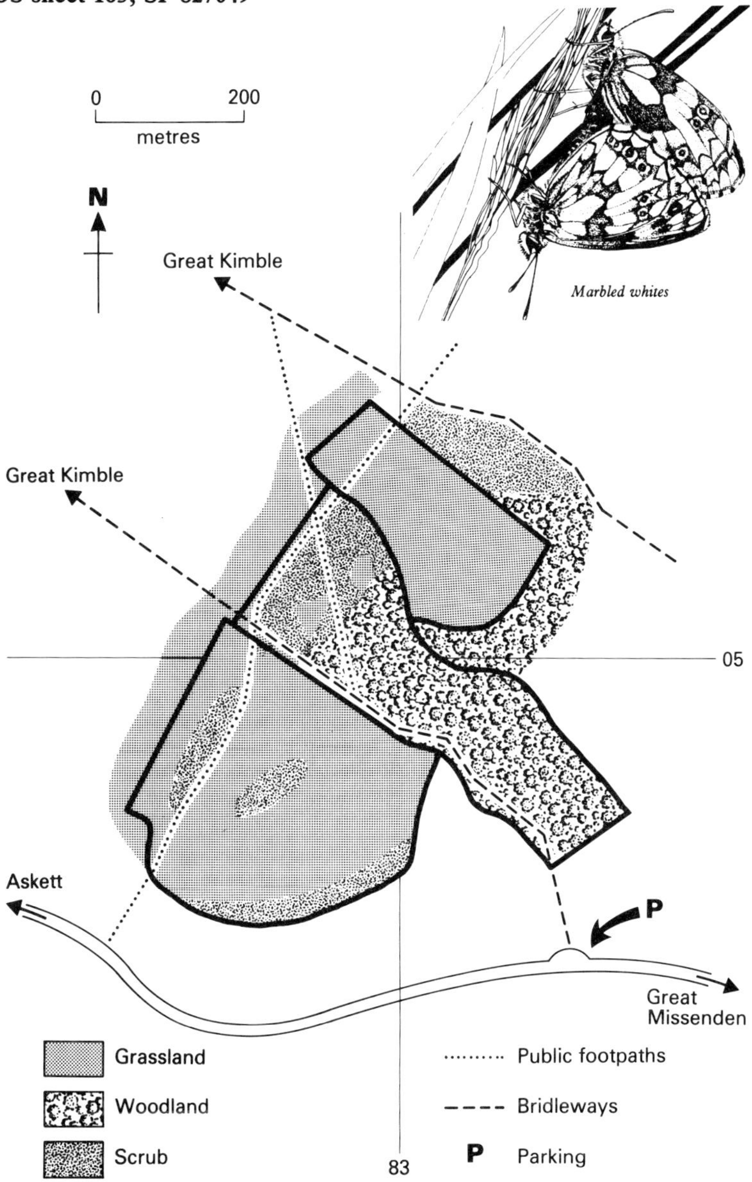

Marbled whites

LITTLE LINFORD WOOD

OS sheet 152; SP 832455

Nearest towns Milton Keynes and Newport Pagnell

An ancient woodland, with some felled areas which have been replanted, comprising 42.5 hectares (105 acres). It has been owned by BBONT since 1986.

Location
From Newport Pagnell take the B526 north; just before Gayhurst turn left towards Haversham. After ½ mile turn right through a farm gate and follow the rough track under the M1 motorway, past Dairy Farm to the car park at Little Linford Wood.

Access
Open to the public on foot along all rides and bridleways. Swan's Way, a long distance bridleway from Salcey Forest to Goring-on-Thames, runs through the wood. Horses must be kept to the bridleways. Dogs must be kept on leads.

Description
This wood is marked on a map dated 1766 and so it is likely that it has never been cleared for agriculture, in other words it is an ancient wood. From medieval times and often much earlier, these old woodlands were being managed without much replanting, the villagers exploiting the fact that trees do not have a fixed life span. Trees and shrubs were regularly coppiced for a reliable supply of small wood, while certain saplings were left (or planted) as 'standards' to grow tall, to be felled each in their prime for timber. The woodland must always have been rich in Spring flowers. In fact the coppice cuts, regularly opening the woodland floor to the light, positively encouraged the bluebells and other flowers.

The recent history of Little Linford Wood is more severe, however. It seems that it was felled and replanted about 100 years ago. Just before it passed into BBONT's care, about a third of the area was clear felled, and some 70 or 80 mature oaks also taken from the northern area. Coppice regrowth in the felled area is taking place however, and with some replanting, the wood will recover in due course.

The principal trees are oak, ash and field maple, and there is also some wild cherry. Some sycamore is present, but this must be controlled. So far about 130 different species of flowering shrubs, woodland flowers and other plants have been recorded, including 23 found only in ancient woodland, such as the local herb-Paris (flowering 5–6).

Management
A great deal of work has already been carried out but much remains to be done. Replanting and work on the rides were carried out during 1987/88; the open rides will be managed to create conditions suitable for a wide range of species and to act as routes for timber extraction. A car park and new pond have been created at the entrance. Some coppicing has been carried out and a start made on a nature trail.

LITTLE LINFORD WOOD

OS sheet 152; SP 832455

Wood-sorrel

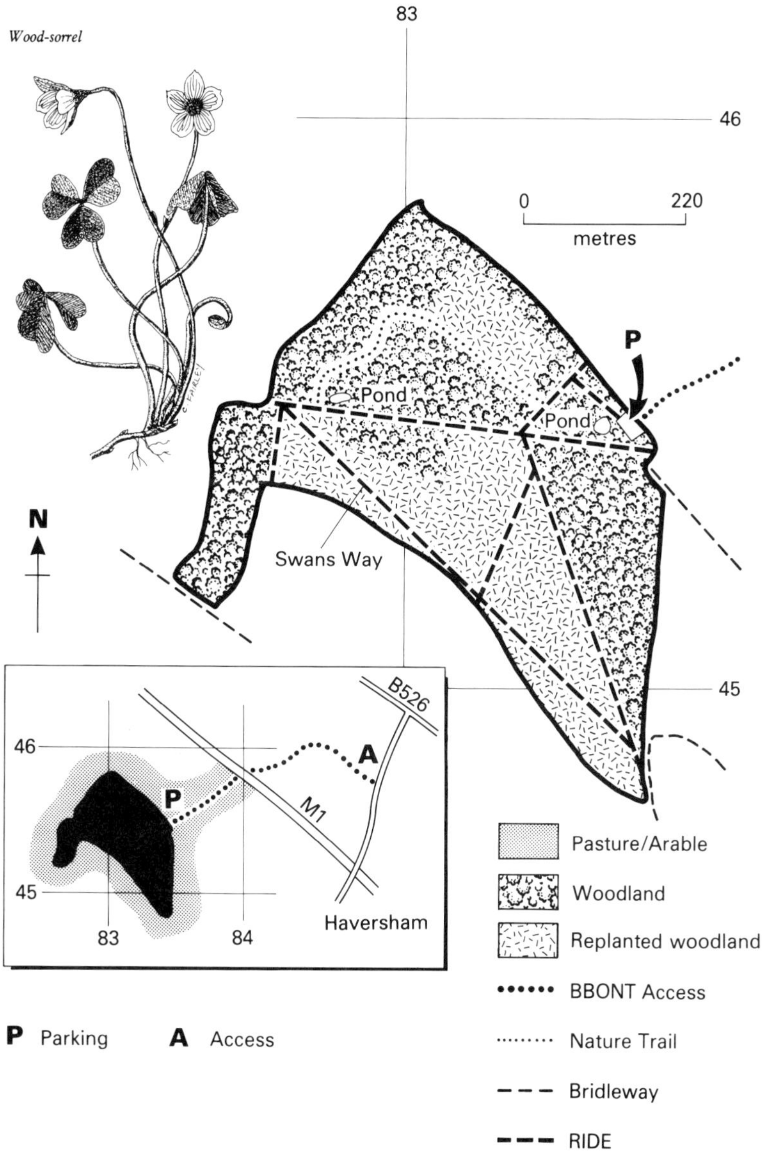

LONG GROVE WOOD

OS sheet 175; SU 963917

Nearest towns Beaconsfield and Amersham

A fragment of old mixed deciduous woodland, 1.6 hectares (4 acres) in area. BBONT was given the freehold in 1964; it is one of our earliest reserves.

Location
From the roundabout at the junction of the A40 and A355 in Beaconsfield, take the A355 north for Amersham. After about 1 mile, take the right turn (Long Bottom Lane) to Seer Green and Chalfont St Peter. Take the second left along this road into Seer Green and turn immediately left into Long Grove. After 400 yards park on the verge on the right-hand side of the road where a public footpath leads off to the north-east. The reserve may be reached by walking down this path for about 50 yards. [The reserve is where the Gr of Seer Green is printed on OS map 175.]

Access
Open to the public.

Description
This is a remnant of old mixed deciduous woodland which once covered much of the Chilterns. The soil here is relatively poor – the wood is rooted into clay-with-flints with some gravelly intrusions. It contains a mixture of oak, beech, hornbeam, field maple and birch. Wild cherry would also be typical, but now there is only a solitary tree. Many ancient and gnarled stools testify to prolonged coppicing of the beech in the past, but some of these stools have died in recent years and a number are moribund through lack of light. The re-establishment of coppicing of the old stools and felling selected trees has encouraged those with sufficiently vigorous root systems to regenerate. This opening up of parts of the wood has also resulted in the appearance of many rowan saplings.

There was further, unplanned clearance in the Great Storm of October 1987, and this has brought light to the ground to the benefit of the flowers which include bluebell (flowering 4–6), and common cow-wheat (5–9). This last is worth looking out for – it is partially parasitic on other plants and is an indicator of old woodland.

It is interesting to see these old beech coppice stools in a wood of this kind. Beech is known for creating its own tall, clean-trunked woodland (the inspiration for the architecture of Gothic cathedrals, some believe). However on poorer soils it tends to grow rather ungainly, and (its timber value lost) it is often coppiced as here or pollarded (as at the famous Burnham Beeches).

Management
The plan is to cut the old stools to help revive them, while other younger shoots will be cut to establish new coppice stools. Amongst them, selected young trees will be left as 'standards' to grow tall. There are also plans to plant wild cherry into the wood.

LONG GROVE WOOD

OS sheet 175; SU 963917

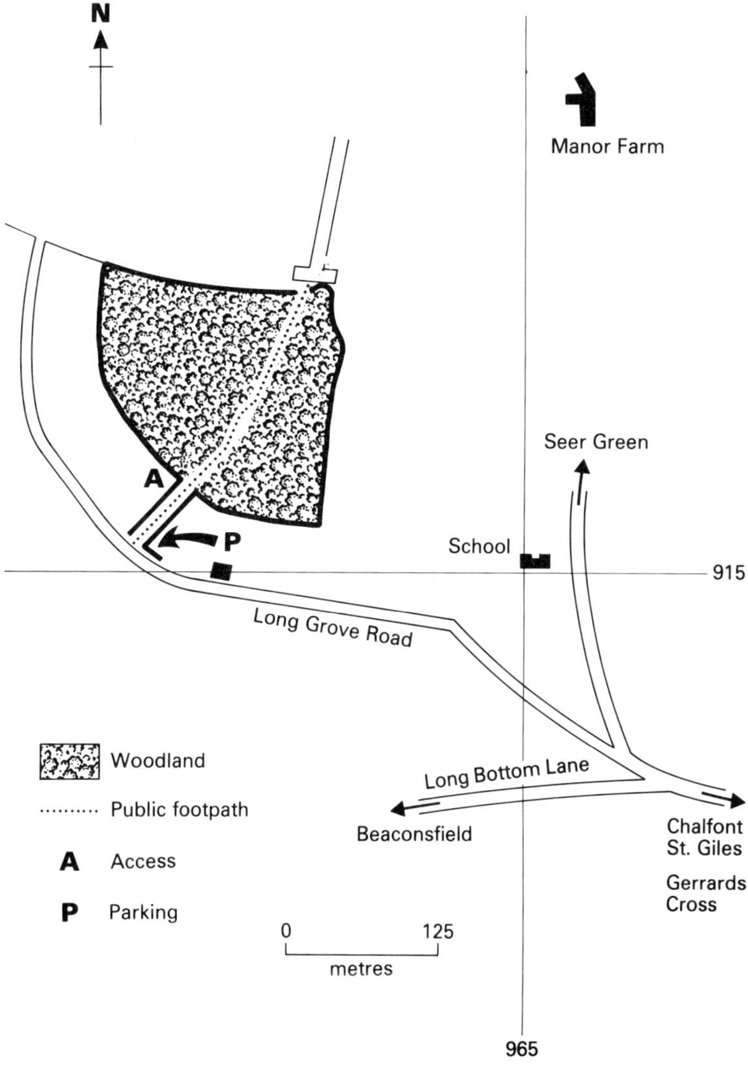

LONG HERDON MEADOW

OS sheet 165; SP 648204

Nearest town Bicester

A small remnant of ancient meadowland, 4.4 hectares (11 acres) in extent. It is an SSSI, bought by BBONT in 1981.

Location
Leave Bicester on the A41 towards Aylesbury and take the second road, signposted to Marsh Gibbon. After ¾ mile, park on the verge just past Grange Farm. Enter through the gate and walk down the right hand hedge to the reserve, which comprises the last two fields before the River Ray.

Access
Open to the public.

Description
These two fields of 2 and 9 acres form a small relic of traditional farming. In olden days, hay was an essential crop, keeping the horses and other livestock through the Winter when the grass in the fields failed. Each village had its hay meadows, often along streams, for grass grows better on damp soil. In recent decades over 90% of such meadows have been 'improved', artificially fertilised to promote better growth of grass or drained and ploughed for cereals or for rye grass.

There are traces of ridge and furrow in the smaller field, evidence of ploughing in the distant past, but both have escaped modern improvement. Indeed, the larger field still lives up to the description 'flood meadow' by being periodically flooded in Winter, an increasingly unusual sight in our over-ordered countryside today.

The fields carry a rich assemblage of grasses, sedges, and flowers such as meadow thistle (flowering 6–8), common meadow-rue (7–8) and great burnet (6–9) and there is plentiful invertebrate life. Snipe and other birds typical of flood meadows might be expected to breed here.

Wet meadows such as this are important for bird life in every season of the year. The reserve lies within the area embracing the headwaters of the River Ray, an area which still holds large wintering flocks of waders such as lapwings. In common with many other species, however, they respond to the weather and in bleak and freezing times will make quite lengthy journeys to find suitable feeding grounds. With delicate bills and feeding by probing for worms and other invertebrates, they and other waders must find soft unfrozen soil. British flocks tend to move to the milder south west or south (and even cross the Channel) in hard spells.

At Long Herdon we can also expect to see flocks of golden plover in Winter. Our British resident birds breed on the moors but come to lowland feeding grounds such as this in Winter. However, golden plovers from Iceland and Scandinavia might also be seen here in Winter.

Management
The meadows are cut for hay late in the season and the aftermath is grazed.

LONG HERDON MEADOW

OS sheet 165; SP 648204

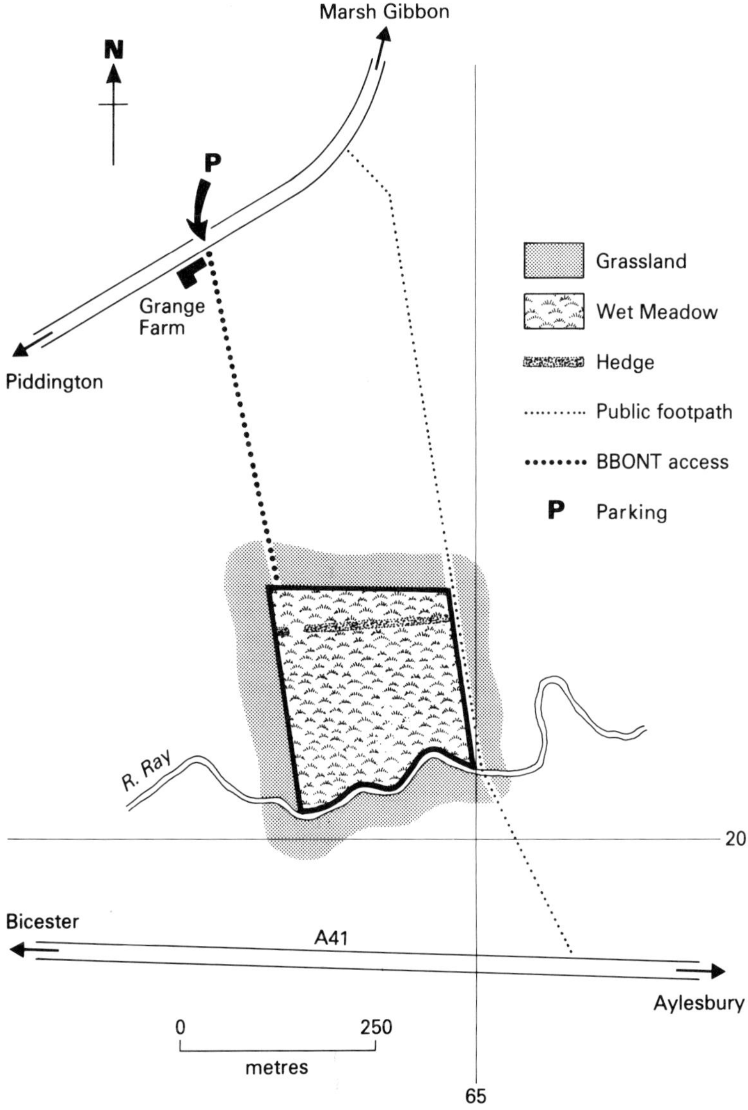

MILLFIELD WOOD

OS sheet 165; SU 870954

Nearest town High Wycombe

An example of a Chiltern beechwood, growing on chalk, totalling 7.3 hectares (19 acres). It is an SSSI, bought by BBONT in 1983.

Location
Take the A4128 from High Wycombe up the Hughenden valley and turn left up the drive to Hughenden Manor. Park in the car park near the church and walk back down the drive and straight across the road to the bridleway. This will take you up the hill, to the reserve, which is on the right. Take care crossing the busy A4128.

Access
Open to the public; there is a public footpath through the reserve.

Description
This is a rare example of a semi-natural Chiltern beechwood growing on chalk – an open beechwood, a 'high forest' in profile with tall trees, mainly beech but also with some ash and wild cherry. There is at a lower level a rather open 'understorey' of coppiced hazel, holly and field maple, and (towards the lighter edges of the wood) whitebeam.

Normally a beechwood is rather bare with few flowers in Spring, but here we can find a number, many of them indicators of ancient woodland, such as goldilocks buttercup (flowering 4–6), wood anemone (4–6) and herb-Paris (5–7), a plant worth looking out for, its leaves held in a hand of four atop the stem. Lily-of-the-valley also flowers here (5–6) – familiar from gardens, it is a native plant. There are bluebells (4–6) and scatters of sanicle (5–8).

The wood is well populated by birds. One to listen for from late March to July is the chiffchaff. Muntjac deer also live in the wood.

Beechwoods such as this can be something of a puzzle. After the last Ice Age, trees from Europe (of which Britain was still part) began to invade to create a wildwood. Beech was the last to arrive before the Channel was cut – it was the last of our 'native' trees. It can grow to overtop and shade out the oak, but Britain is on the fringe of its natural range and it cannot set viable seed far to the north (in fact, all beech trees north of Derby may have been planted). Beech is valued and planted for its timber, and documents show that many Chiltern beechwoods were open sheepwalks 400 years ago. On the other hand, Buckinghamshire could well have gained its name from natural woodlands of beech – the Old English name of this tree was 'bece', which easily becomes 'buck'.

As well as coppice wood, timber was almost certainly extracted from this wood in the past. Nevertheless it now has a natural aspect – hence the description 'semi-natural'.

Management
Clearing and maintaining the footpaths, removal of sycamore, horse chestnut and other invasive species and creation of open clearings to encourage the ground flowers and butterflies.

MILLFIELD WOOD

OS sheet 165; SU 870954

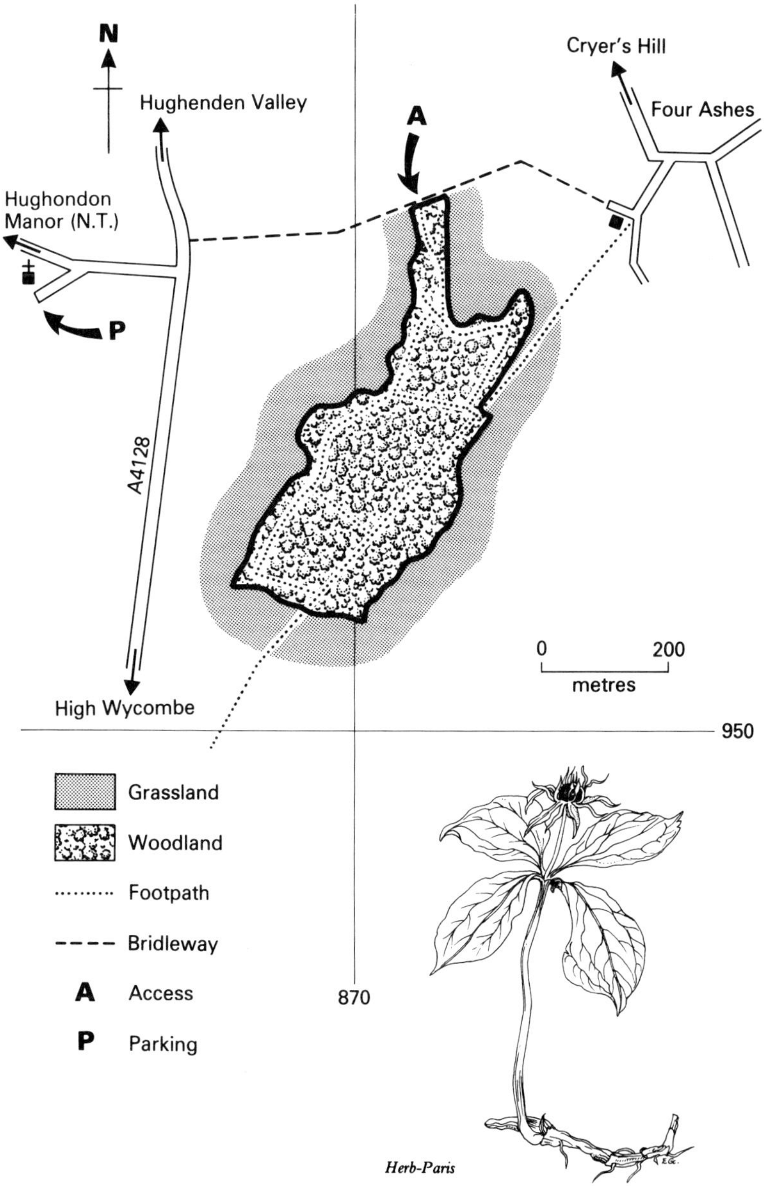

Herb-Paris

OAKLEY PARISH HEDGE

OS sheets 164 & 165; SP 615122

Nearest towns Oxford and Thame

An ancient hedgerow, managed by BBONT by agreement with the owners, the Brill and Oakley Relief in Need Charity. The land area is about 0.2 hectares (½ acre). The agreement began in 1984.

Location
From Headington roundabout on the north-east side of Oxford travel northwards past the crematorium and turn right onto the B4027. Turn left into Stanton St John and first left in the village. Continue to the road junction at Menmarsh and then fork left and proceed for about 1¼ miles. The hedge is on your right, beginning at the north corner of Oakley Wood and ending at the field gate, before the first house on the right.

Access
The hedge is along the roadside.

Description
The reserve consists of the hedge behind the south-east roadside verge. It is an historic hedge and probably the oldest thing in the parish, and it has an interesting story to tell.

First of all, there is something unexpected to notice, for while a ditch normally lies alongside a road and the hedge beyond it (as is the case on the opposite side of the road), here they are reversed, with Oakley Parish Hedge lying between road and ditch. Moreover it grows on a substantial bank well above the field behind. A massive bank and ditch such as this suggest an important ancient boundary, which the road later followed.

All the evidence indicates that the boundary was that of the Royal Forest of Bernwood, marking off the Forest from Oakley Wood which once extended up to it. It bounded an area of the Royal Forest marked as 'Stonehurst' on a 1590 map. On disafforestation of the Forest in 1632, in lieu of commonland lost, this area was given to the commoners, and on an 18th century map it appears as 'Poor Folks Pasture'. Today it is known as Pasture Farm. The Brill and Oakley Relief in Need Charity probably dates from the 17th century, and might be in some way a successor to that grant of land. Interest from the sale of Pasture Farm a few decades ago is still used for the benefit of the people of Oakley and Brill. The hedge is now part of the parish boundary between Oakley and Boarstall.

At some time Oakley Wood was also cleared back from the boundary. Thus, the hedge we see today may have its origins in part or whole from a strip of woodland or scrub left when open ground was cleared on each side. The handy method of dating field hedges (see page 106) does not apply here; nevertheless it is interesting that 17 woody species have been counted.

Management
Oakley Parish Hedge appears always to have been managed by coppicing, and that tradition is being continued. There is yearly coppicing of the hedge in sections, but leaving the larger trees as standards.

OAKLEY PARISH HEDGE

OS sheets 164 & 165; SP 615122

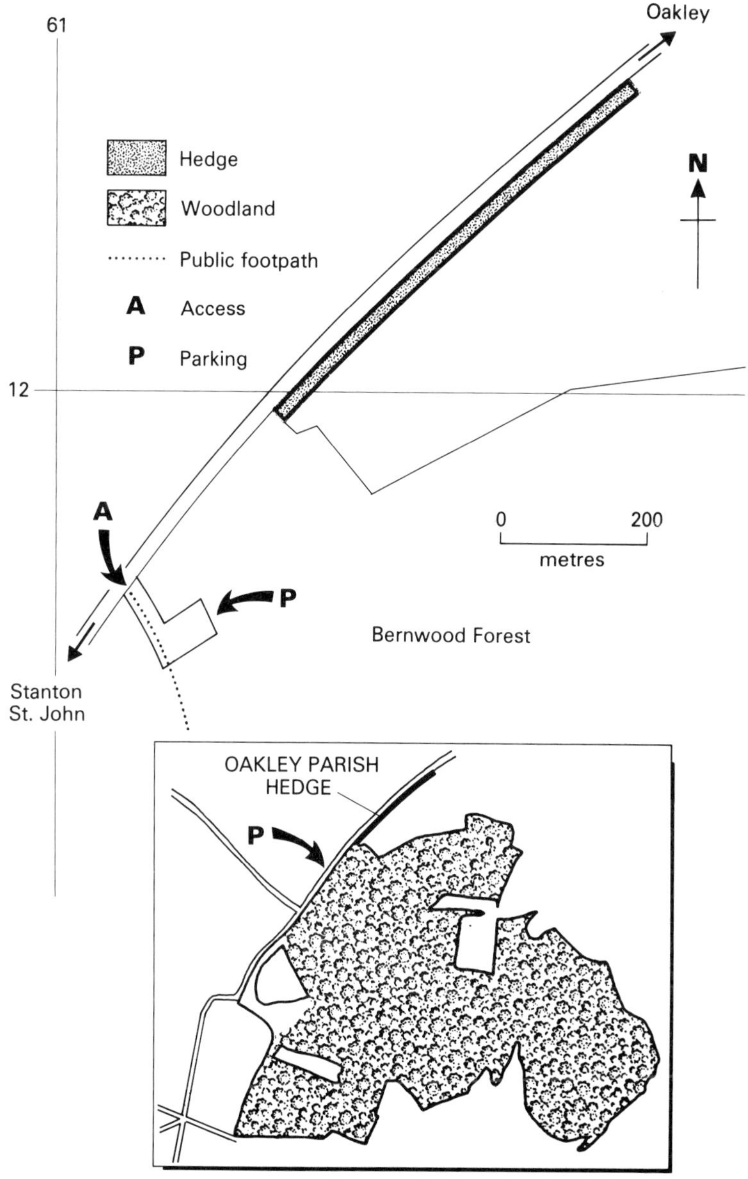

OXFORD LANE

OS sheet 165; SP 698189

Nearest towns Bicester and Aylesbury

An ancient, hedged, green lane, 2.1 hectares (5¼ acres) in area. BBONT manages it by agreement with the owners dated 1981. It forms part of the Ham Home-cum-Hamgreen Woods SSSI.

Location
Leave Aylesbury or Bicester on the A41. If driving from Aylesbury, go through Waddesdon and Woodham. Park in the layby on the right, half a mile past Woodham, at the corner of Kingswood at SP 695186. The reserve is the old green lane shown on the map opposite.

Access
Open to BBONT members only. The shooting is let and dogs are not allowed.

Description
This is an ancient green lane, bordered on the west by woodland and on the east by pasture and arable fields. The hedges on each side are mainly of hawthorn and blackthorn (sloe) which are spreading inwards.

There is also an open area of grassland towards the northern end where the edge of the woodland retreats.

This grassy area and the grassy parts of the lane carry a wide variety of grasses and wildflowers. Sneezewort (flowering 7–8) is to be seen, as are agrimony (6–9), betony (6–9) and devils'-bit scabious (6–10).

The greatest interest of the reserve, however, lies with its butterflies. Over 20 species have been recorded. This is a good place to look out for the black hairstreak butterfly – which is described on page 68 . The white admiral is also seen; this is a woodland butterfly which lays its eggs on honeysuckle growing in the shade below the trees. It is most often glimpsed in rides and clearings where flowers provide nectar; it visits Oxford Lane from the wood alongside. Other species to look for include the marbled white, a butterfly of open grassland, and the comma. This last has been successful in recent times. A century ago, comma numbers were declining and the butterfly became virtually unknown outside the Wye Valley. Now it is seen widely in southern England, the Midlands and Wales and is even one of the commoner butterflies of suburban gardens. It is one of the latest butterflies on the wing, flying well into October.

There is a small pond at the edge of the wood, as the map shows. Here damselflies and dragonflies are to be seen. The latter often range far from water, however, and may be encountered down the length of the reserve.

Management
BBONT is responsible for maintaining an extraction route through the scrub which can at need be used by timber lorries. This scrub contains a large amount of blackthorn, important for the black hairstreak butterfly, and coppicing has been started to give bushes of various ages.

The grassland also needs managing and we propose to mow it annually.

OXFORD LANE

OS sheet 165; SP 698189

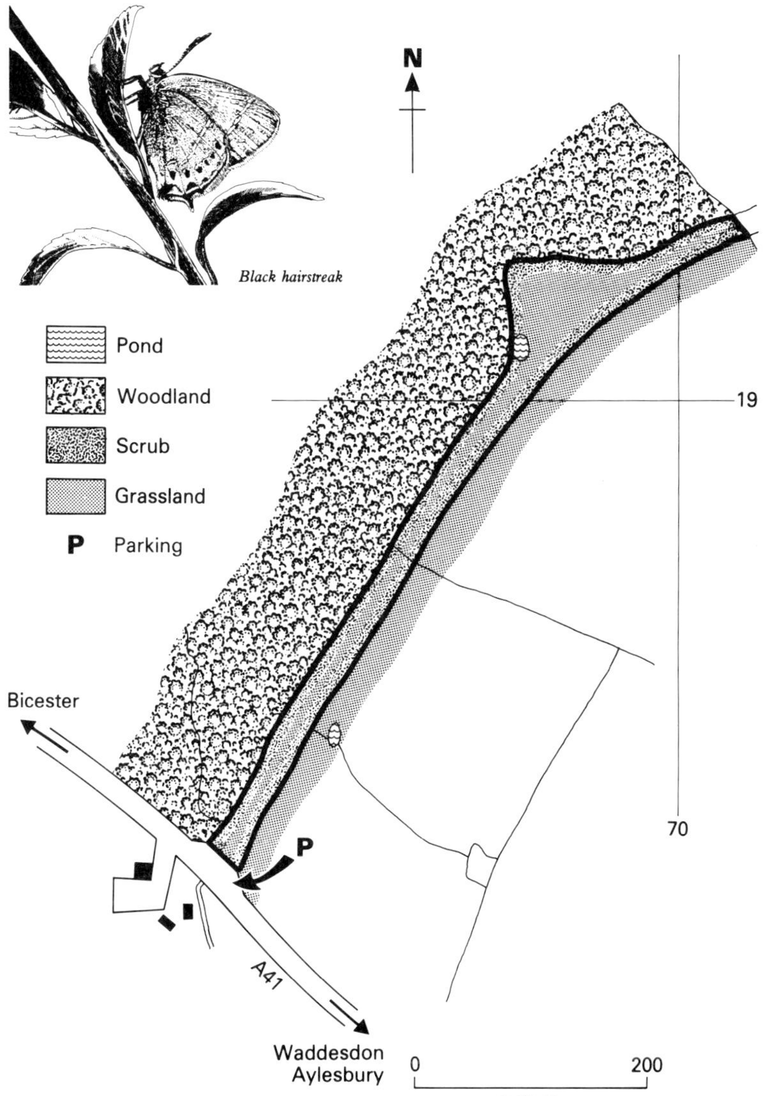

PARK WOOD

OS sheet 165; SU 823987

Nearest town Princes Risborough

A small area of 1.2 hectares (3 acres) of chalk grassland and scrub dominated by whitebeam, on lease from the National Trust since 1969. The reserve forms part of the Bradenham Woods SSSI.

Location
Leave Princes Risborough on the A4010 southwards for 3½ miles to Saunderton, and turn left into Small Dean Lane, a minor road opposite the Golden Cross public house. Cars may be parked in the National Trust car park 1 mile down the lane on the right. Follow the footpath leading from the southern corner of the car park for about 100 yards then take the bridleway which branches off to the right down a steep slope. The reserve entrance is about 200 yards along the bridleway on the right. An easier route is via the public footpath which leads south from Small Dean Lane through the farmyard of Small Dean Farm and running along the foot of the reserve.

Access
Open to the public. Note that from November to March there may be shooting on adjacent land.

Description
The reserve is a mosaic of chalk grassland and scrub on the western edge of Park Wood. Apart from hawthorn and other usual constituents, the scrub contains a large amount of yew and whitebeam. These add visual as well as ecological interest – the yew of course retains its dark green foliage all year round, while the silvery-leaved whitebeams make a prominent and handsome show which is enhanced in early Summer by their white flowers (5–6) and in Autumn by their reddish berries.

The glades, which have been maintained by the rigorous control of the ever-invasive scrub, carry a range of chalk grassland plants including clustered bellflower (6–10), Chiltern gentian (8–9) and common spotted, pyramidal, fragrant and bee orchids. June is the prime flowering month for most orchid species. In all, over 130 plant species have been recorded on this small reserve.

24 species of butterfly have been recorded here; of special interest is a breeding colony of the Duke of Burgundy butterfly. With its speckled appearance this small butterfly resembles the larger woodland fritillary butterflies (it is sometimes known as the Duke of Burgundy fritillary). It is on the wing from mid May for about a month, making short, fast, restless flights. The food plants for its caterpillars are primrose and cowslip, common enough, but the butterfly itself is now rare, with rather local distribution.

Management
This aims at maintaining the attractively varied mosaic of scrub and grassy glades which form the habitat favoured by the Duke of Burgundy.

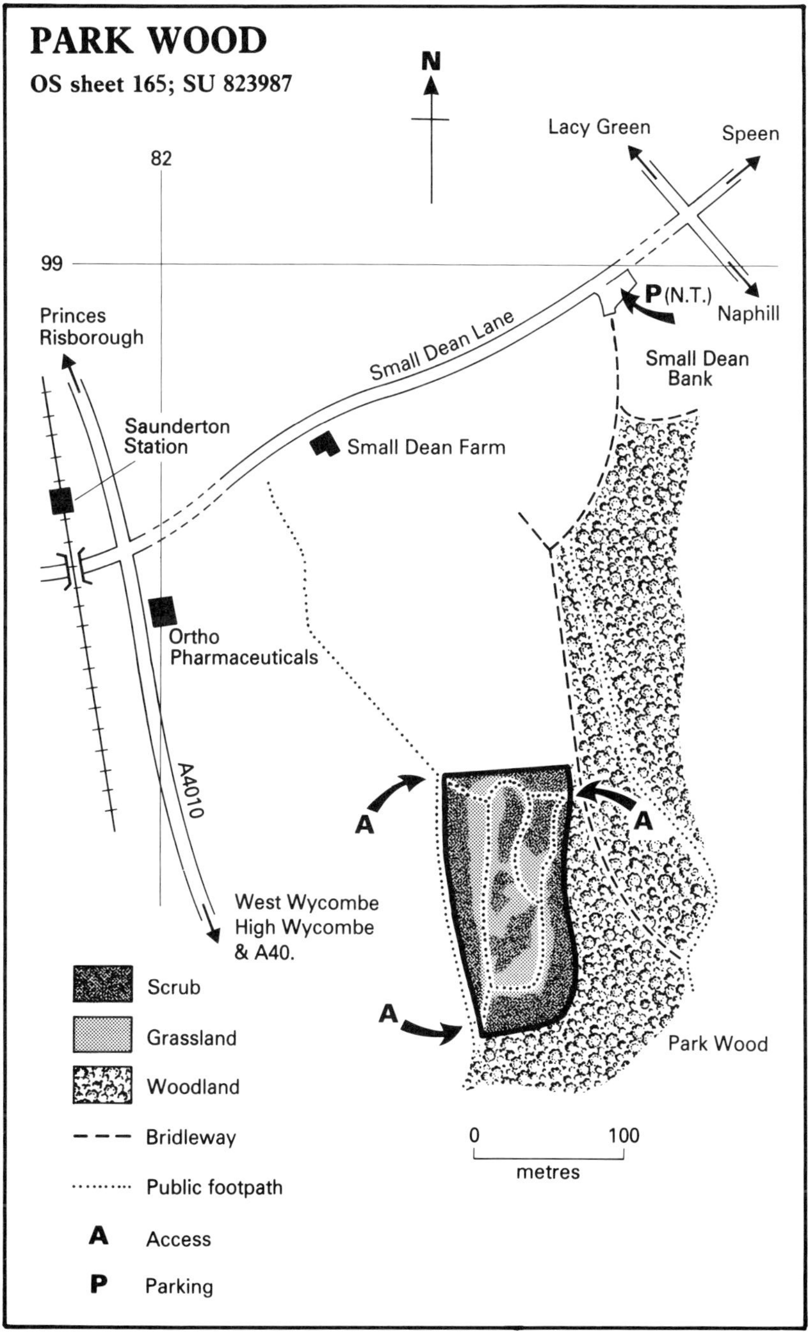
PARK WOOD
OS sheet 165; SU 823987
N
82
99
Lacy Green
Speen
Naphill
P (N.T.)
Princes Risborough
Small Dean Lane
Small Dean Bank
Saunderton Station
Small Dean Farm
Ortho Pharmaceuticals
A4010
A
A
A
West Wycombe
High Wycombe
& A40.
Park Wood
0
100
metres
Scrub
Grassland
Woodland
Bridleway
Public footpath
A Access
P Parking

PILCH FIELD

OS sheets 152 & 165; SP 747322

Nearest towns Buckingham and Winslow

A pair of ancient meadows and a small triangular field, 11.86 hectares (29.3 acres) in area surrounded by hedgerows. The whole was bought by BBONT in 1984. The two meadows are a notified SSSI.

Location
From Buckingham take the A421 towards Milton Keynes for about 3 miles and then take the unclassified road on the right, towards Great Horwood. Park on the verge near the sharp left hand bend; the reserve is through the gate to the right of you and consists of the small triangular field plus the large field in front of it and a second large field to its right.

Access
Open to the public, but a local farmer has a grazing licence and no dogs are allowed when stock are grazing.

Description
The two ancient fields are marked with medieval ridge and furrow. Although they have been drained in part, they have escaped deep modern ploughing and there are still major soil differences which show themselves in the distribution of the wildflowers and wild grasses.

The drier areas carry meadow barley, meadow oat-grass and quaking-grass. This last is one of our most elegant wild grasses, its dainty panicles set dancing by the slightest breeze (hence its name) in June. Amongst the flowers seen are dwarf thistle (flowering 6–9), salad burnet (5–8), cowslip (4–5), dropwort (5–8), green-winged (5–6) and bee (6–7) orchids.

In Spring, handsome marsh-marigold (kingcup) flowers (3–5) mark some of the damper soil. There are many wet 'flushes' fed by water seeping through the soil where rushes and sedges grow, and wetland wildflowers such as meadowsweet (6–9), marsh valerian (5–6) and the early marsh-orchid (5–7). One plant to look out for is carnation sedge, which carries handsome panicles in May and June; its leaves are pale bluish-green, rather like the colour of garden carnation leaves, hence its name.

There are plenty of butterflies, grasshoppers and other insects. Look out for the six-spot burnet moth, which can be seen flying on sunny days from June to August. It flies sluggishly and could be caught even on the wing by a nimble bird; however its rather gaudy black and red colouring is a warning to birds and other predators that it is distasteful.

There is a greal deal of bird life in the bordering hedgerows – warblers nest, and kestrels are often seen. From time to time, green woodpeckers come to feed on the meadows. They have a liking for ants. There are a number of ant hills of the yellow meadow ant to be seen (together with the wildflowers, these are also a sign that the ground has remained free of disturbance for a good while).

Management
The fields are grazed by cattle.

PILCH FIELD

OS sheets 152 & 165; SP 747322

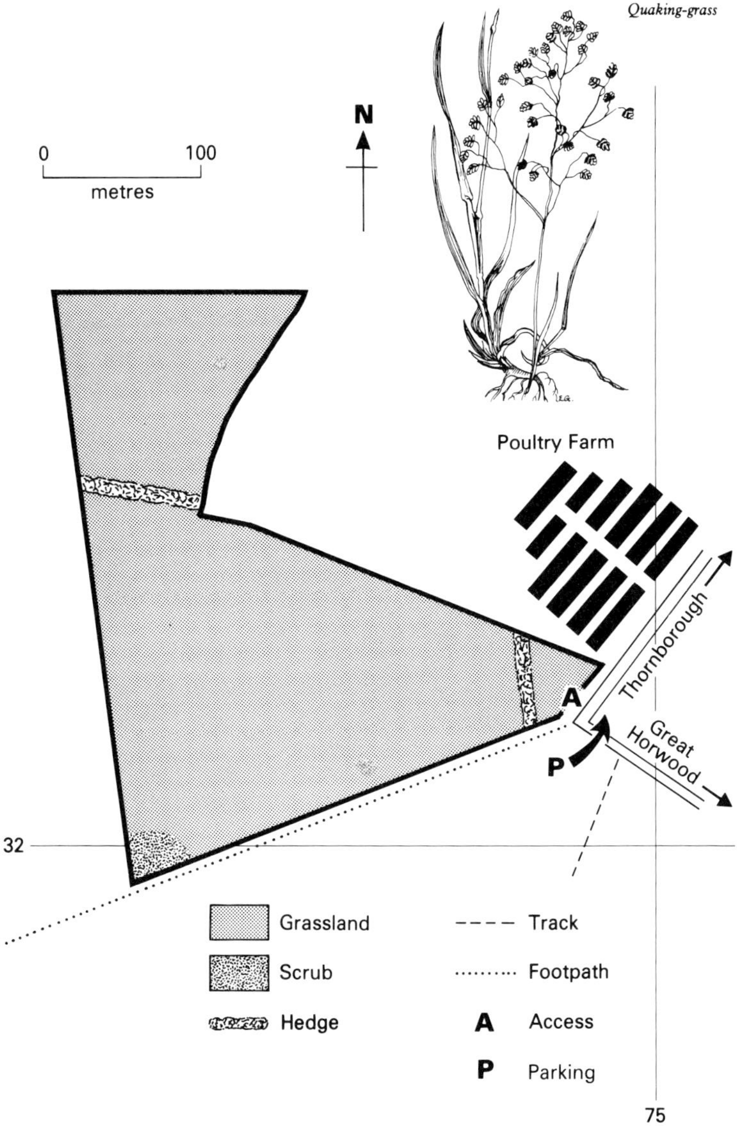

PITSTONE HILL

OS sheet 165; SP 950145

Nearest towns Aylesbury, Tring and Wendover

21.85 hectares (54 acres) of chalk grassland, an SSSI on licence from Buckinghamshire County Council. It is within the Chilterns Area of Outstanding Natural Beauty and makes a fine outing for the nearby views alone. A section of Grimm's Ditch runs through most of its length.

Location
From the roundabout situated on the A41 just west of Tring take the unclassified road north-east past Tring to Tring Wharf. Take the B488 north-east to the T-junction at SP 950155. Turn right towards Aldbury; the car park is on your right on the brow of the hill.

Access
Open to the public. Dogs must be kept on a lead.

Description
The reserve encompasses a steep scarp face, with the chalk exposed for some of its length. The reserve faces southwards to Aldbury Nowers Wood. It is a very good tract of grazed chalk grassland, and although the gentler slopes at the top have had artificial fertilisers applied, the turf of the steeper slopes has escaped this treatment (artificial fertilisation encourages the coarser grasses which then grow to overwhelm the slighter wildflowers).

For the botanist, the steep slopes carry most of the classic chalkland flowers. Prime amongst them is the pasqueflower, its flowering (4–6) may coincide with Easter, presumably the origin of its name.

As interesting as the flowers are the butterflies linked to them. The instinct of the female is to lay her eggs on a particular species of plant on which the caterpillar will feed, and sometimes she will choose no other. Colonies of weakly flying butterflies may thus be quite localised, found only where those plants grow. The chalkhill blue butterfly is one example linked to the horseshoe vetch which is seen flowering on this reserve (5–7), while the small blue chooses kidney vetch (also seen here 6–9). The common rock-rose (5–9) is the choice of the brown argus butterflies.

There are numerous ant hills made by the yellow ant; these can be of considerable age, and are worth protecting on any nature reserve where they occur. The worker ants are rarely seen, since they forage mainly below ground. From time to time in Summer, however, perhaps in response to such things as air temperature and static electricity (a sign of dry air), winged males and females swarm out in a nuptial flight, after which the males die and the females live on to become egg-laying 'queens'.

Skylarks, meadow pipits and other birds sing here, on what is a truly delightful and interesting site.

Management
The reserve is grazed by sheep. Scrub has been removed during the past few years and needs to be kept under control.

PITSTONE HILL

OS sheet 165; SP 950145

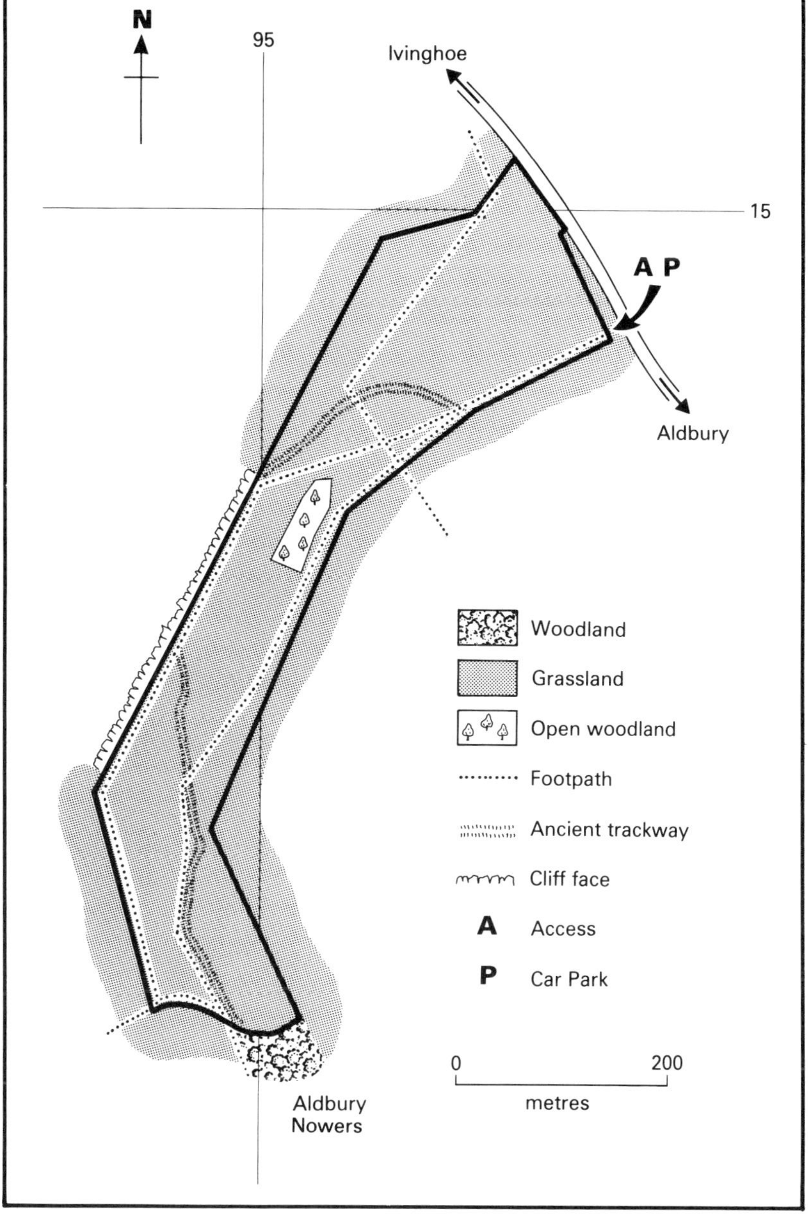

RUSHBEDS WOOD and LAPLAND FARM

OS sheets 164 & 165; SP 668157

Nearest towns Aylesbury and Bicester

Of SSSI status, a woodland of 45 hectares (111 acres) bought in 1983, plus about 6 acres of meadow. In 1985, BBONT also purchased Lapland Farm, 11 hectares (27 acres).

Location
From Aylesbury take the A41 towards Bicester. At Kingswood turn left along the unclassified road south-westwards to the T-junction. To visit Lapland Farm drive straight over and up the track in front, cross the railway bridge and park on hardstanding without obstructing the gates. To visit Rushbeds Wood turn left at the T-junction and just short of the next T-junction turn right up a track, cross the railway bridge and park on the left.

Access
The wood is open to the public along the rides. A public footpath crosses Lapland Farm; please keep to it during the growing season and keep dogs on a lead when sheep are in the fields.

Description
Rushbeds Wood is an ancient woodland which appears on a map of 1590. It is one of the remaining woods of the ancient Forest of Bernwood, and has probably existed as woodland since the return of trees after the last Ice Age. Over 200 species of flowering plant, 50 fungi, and 60 bryophytes (mosses and liverworts) have been recorded. The butterflies include white admiral, purple emperor and a colony of black hairstreak.

The large amount of dead wood provides good habitat for beetles and other invertebrates, which in turn attract numerous woodpeckers. Muntjac deer are common in the wood and foxes are seen frequently.

The two long narrow meadows to the south are sheltered and in Summer many butterflies are seen feeding on the flowers here. Some of these wildflowers are characteristic of old meadows (and there are more in the two old fields of Lapland Farm). The wet patches contain flowers found nowhere else on the reserve. The hedge bounding the road has 19 species of trees and shrub altogether, and the standard hedge-dating method (average number of species in 30 yards = age of the hedge in centuries) suggests that it is at least 500 years old. The hedges are good for birds and butterflies and there is an interesting pond. The drove road north of the fields is also on the 1590 map; drove roads, used for moving livestock, are wider than green lanes.

Management
Most of the trees were felled in the late 1940's and coppicing ceased soon after. Part of the wood is being coppiced again and most of the remainder will be left to grow gradually into a 'wildwood'. Rides have been widened to let in light and encourage more wildflowers and insects and several glades have been created for the same purpose. The fields at Lapland Farm will be cut for hay and grazed afterwards, while Rushbeds Meadows will be grazed by cattle.

RUSHBEDS WOOD and LAPLAND FARM

OS sheets 164 & 165; SP 668157

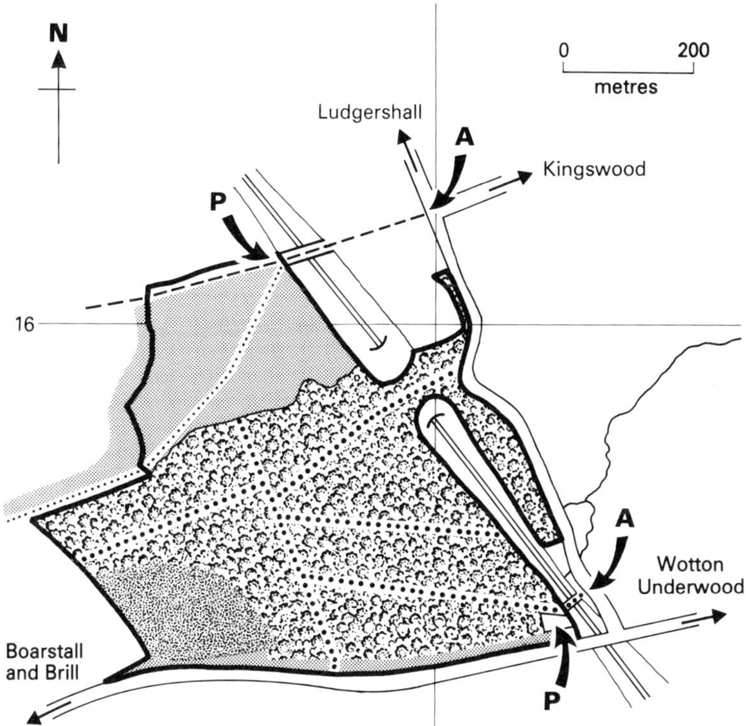

Great spotted woodpecker

STONY STRATFORD

OS sheet 152; SP 785412

Nearest town Milton Keynes

A reinstated gravel working, 23 hectares (57 acres) in area. On lease from Milton Keynes Development Corporation and MK Borough Council since 1982.

Location
From central Milton Keynes take the new A5 north to the roundabout and turn south towards Stony Stratford but then left along the bypass loop road and left at SP 790408 which ends at the car park.

Access
The northern part of the reserve and the path to the bird hide are open to the public. The reserve is a sanctuary for wildfowl, however, so that public access to the fenced area is restricted; a permit may be obtained from the BBONT Office in Oxford.

Description
Parts of the reserve look like an ancient mere, plentifully fringed with reeds and with a richly varied population of wild geese, ducks and waders – the kind of place where Sir Peter Scott might like to prop his easel. Yet a photograph taken a few years ago shows that the site used to be an open meadow where redshank nested. However, under the grass lay deposits of good quality gravel, ideal for the new bypass to be built nearby. It was at this point that the planners of MK Development Corporation made an imaginative leap and envisaged a prize nature reserve within the boundaries of their new city. They realised that the gravel extraction could be seen not as a threat to wildlife but as an opportunity. If it was excavated to a prearranged design, the eventual flooding would produce a purpose-built wetland, with lagoons both shallow and deep and islands where wildfowl could nest safe from rats and foxes (and vandals of the human kind).

A deal was struck – the road builders were to have this local gravel and save on transport costs, while establishing an endowment fund to yield an income to help maintain the new reserve. The land was then leased to BBONT. Even before the last of the machinery left the site, waders, duck, geese, swans, grebes, moorhens and coots were nesting and they still do.

BBONT's reserve warden and volunteers were in from the start. An early diary entry shows that on one single day no fewer than 700 bur-reeds, 1000 flowering rushes, 25 bulrushes and 100 yellow iris were planted by hand. The resulting look of instant maturity not only satisfied the birdlife, but also local residents who were then more willing to lend their moral support. Vandalism is still a problem, however – the hide has twice been burnt down, but restored afterwards.

Management
Gravel must be brought in to maintain the islands. The public open space is mown, and grazed afterwards, other areas are also grazed.

STONY STRATFORD

OS sheet 152; SP 785412

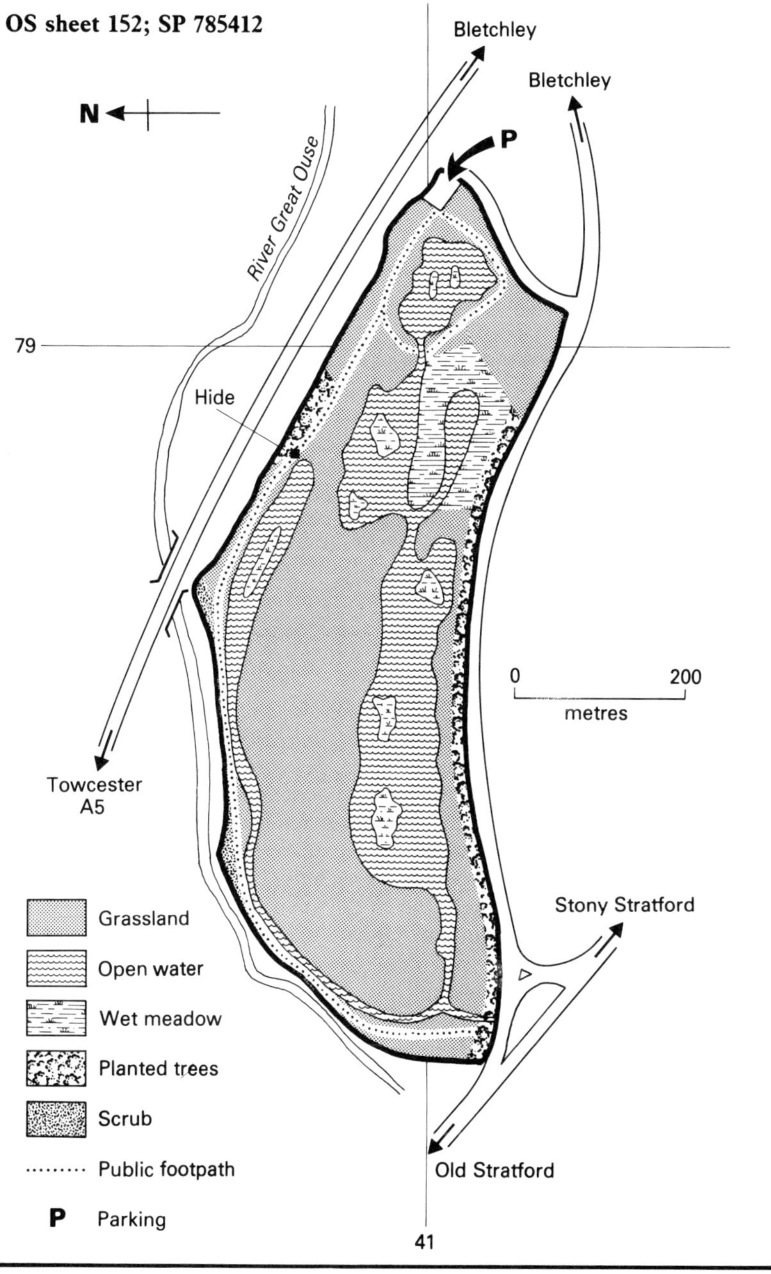

THE STRAITS, Salcey Forest

OS sheet 152; SP 812513

Nearest town Newport Pagnell

An area of woodland totalling 13.6 hectares (34 acres). It is an SSSI managed jointly by BBONT and the Northamptonshire Wildlife Trust (NWT) under an agreement with the Forestry Commission dated 1971.

Location
From Newport Pagnell, travel north-westwards along the B526, go through Stoke Goldington and after 2 miles turn left along the unclassified road towards Hartwell. After a quarter of a mile you enter Salcey Forest. Park in a layby on the left, 600 yards after entering the Forest. Cross the road and walk up the Forestry Commission ride northwards. The reserve is on your right after 300 yards.

Access
Salcey Forest is open to the public along the main Forestry Commission rides, but entry to the reserve itself is restricted to members of BBONT and NWT. Care should be taken not to trample the rare plants of the rides, especially when muddy underfoot.

Description
The names of Great and Little Straits which form the reserve go back to medieval times when Salcey Forest was one of the royal hunting Forests. Although much of Salcey Forest has now been converted to conifer or mixed conifer-broadleaved stands, the Straits carry native oak planted in 1847 after the previous timber was felled. Other native trees now grow amongst the oaks, including ash and field maple. Sycamore, sweet chestnut and turkey oak (with more sharply pointed leaves than native oaks and acorns in scaly cups) are introduced species which also grow here.

The shrub layer below the trees is mainly coppice growth of hazel and field maple with some hawthorn, including Midland hawthorn. This last is not restricted to the Midlands but (unlike the more familiar common hawthorn of the hedges) is likely to be found in old woodland. It can be recognised by its blunter leaves. It flowers about a fortnight in advance of the common hawthorn and now provides us with 'may' blossom to see in the month of May (until the calendar was adjusted two centuries ago, and 11 days were lost, the common hawthorn fulfilled that role).

Below the shrubs are a host of wildflowers, from the familiar bluebell (flowering 4–6) and wood-sorrel (4–5) to unusual species such as herb-Paris (5–7) and pignut (5–6). Ferns are common in many areas and sedges are a special feature; clumps of pendulous sedge may reach 6ft high along the rides.

Salcey is also renowned for its woodland butterflies. 52 species of birds nest, including the nightingale, and there are fallow as well as muntjac deer.

Management
Ideally the oaks should be managed to grow to maturity, drop boughs to rot for the benefit of beetles and other invertebrates, seed themselves . . . to create, in time, a true wildwood.

THE STRAITS, Salcey Forest

OS sheet 152; SP 812513

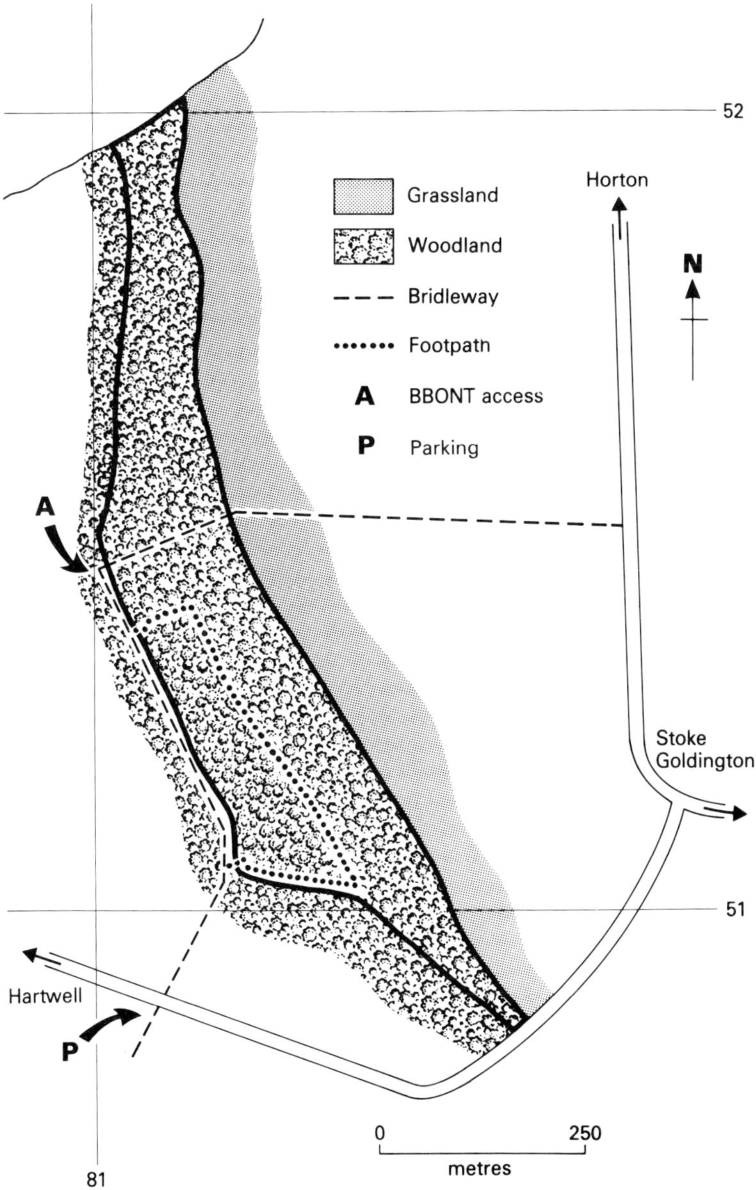

WESTON TURVILLE RESERVOIR

OS sheet 165; SP 862096

Nearest towns Aylesbury and Wendover

An old reservoir with 4.8 hectares (12 acres), an SSSI managed by BBONT by agreement with the British Waterways Board since 1966.

Location
From Wendover take the A413 north; turn right after one mile opposite the Marquis of Granby public house. Park in the layby after half a mile, and the reserve is on your right.

Access
Open to the public along the perimeter path only.

Description
The open water is leased by the British Waterways Authority to the local sailing and angling clubs. BBONT's agreement covers only the surrounding land and reedbeds, but BBONT has, since 1963, had shooting rights over the whole area. These rights are not exercised.

The reservoir was constructed in 1795 to supply water to an arm of the Grand Union Canal, and over the years a good many wetland plants have colonised its margins. There is now an extensive reedbed, and a marshy fen. In addition to the reed (its tassels notable in late Summer), plants include the lesser bulrush, early marsh-orchid (flowering 5–7) and orange balsam (6–8). This is less familiar than the purplish flowered Indian balsam (neither of them are native species).

There are good breeding colonies of reed warbler. This bird is perhaps less likely to be seen than heard: its song is a flow of churr-churr-churr . . . chirruc-chirruc-chirruc notes. It makes a beautifully woven cup nest, slung between the reed stems, deep enough to prevent eggs or chicks being tossed out when the reeds sway in the wind. The site might attract cuckoos in Spring, for the reed warbler is one of the cuckoo's main hosts. The reserve is also the only regular breeding site in Buckinghamshire of the water rail. This bird, too, is much more often heard than seen with a call sounding like a piglet squealing. It can regularly be heard all year round.

The reserve is renowned as an Autumn roost area for swallows which gather into large flocks before leaving (suddenly, when wind and weather are right) on migration. Starlings also roost here in some numbers in Autumn.

The reserve is a good place to observe wildfowl, both Wintering and on passage to and from breeding grounds to the north of Britain. In April 1988, for example, two female eider duck were seen here – the first record of this species in Buckinghamshire.

Management
Attempts have been made to remove sycamore from the woodland area, and the marsh area is kept free of sallow scrub.

WESTON TURVILLE RESERVOIR

OS sheet 165; SP 862096

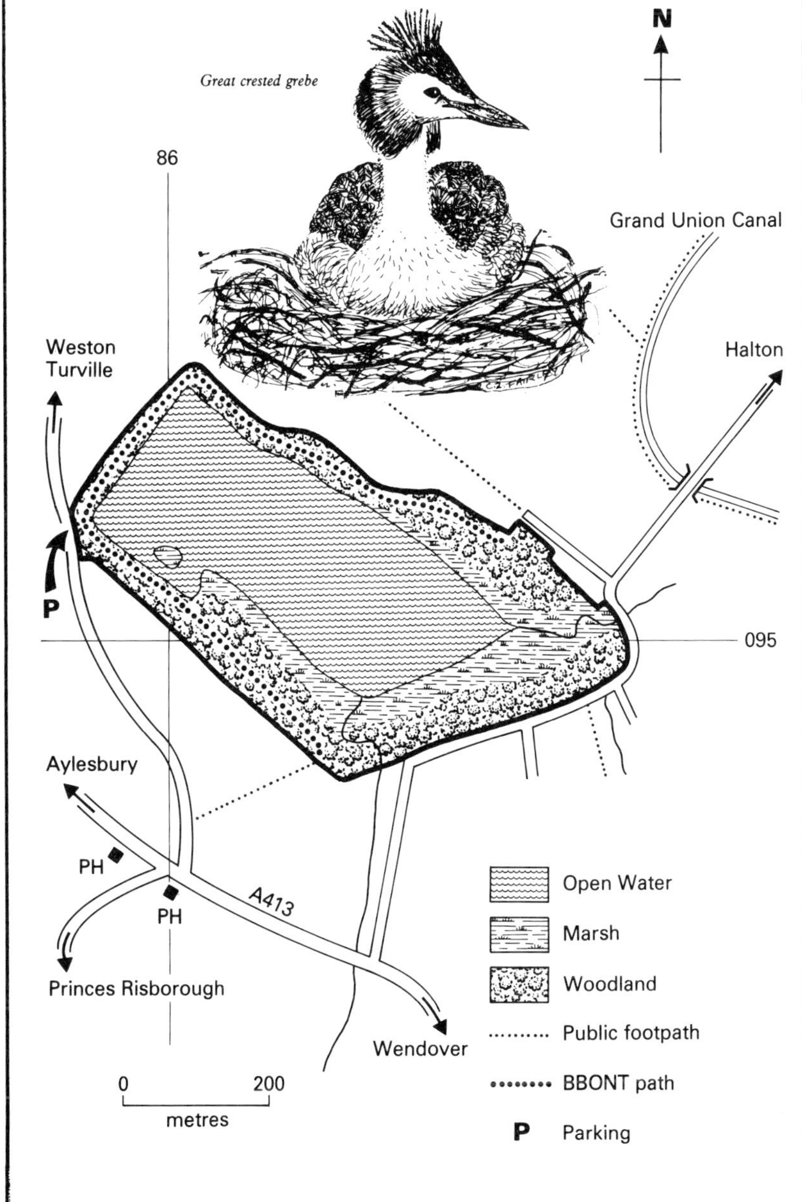

WINDSOR HILL

OS sheet 165; SP 826027

Nearest town Princes Risborough

An area of beech woodland and chalk grassland, 6.8 hectares (17 acres) in area. It is part of the Windsor Hill SSSI, and has been managed by BBONT by agreement with the owners since 1973.

Location
From Princes Risborough follow the A4010 for 1 mile and turn right at SP 814043 up Whiteleaf Hill for 1½ miles. Go past the cottages (on your left) called Green Hailey and park on the verge at the end of the field. Walk along the bridleway at the edge of the field to the wood; turn left then right and follow the broad track gently downhill for about ¼ mile. Here, on the left, an area of chalk grassland is reached. This, and the beech woodland surrounding it, are parts of the reserve and there are other patches of grassland less easily accessible.

Access
Open to BBONT members only. The shooting is let by the owner and visitors should not enter if a shoot is in progress.

Description
This reserve is partly Chiltern beech woodland and partly old chalk grassland carrying wildflowers such as common rock-rose (flowering 5–9), common valerian (6–8) and wild candytuft (7–8) – this last often does well on ground grazed by rabbits.

There is some juniper scattered on more open ground. The grassland was at one time much more extensive but scrub and trees have invaded and the juniper has succumbed to the dense shade they have created.

The Black Hedge which runs through the reserve was mentioned in a Charter of 903 AD and may have been 200 years old then. It bounds Monks Risborough. Although the handy method of dating field hedges (see page 106 and the notes on OAKLEY PARISH HEDGE) does not apply to old boundary hedges such as this, which possibly originated as a strip of woodland left when land was cleared on both sides, it is interesting that a count of the shrubs in it gave an average of 12.4 different species per 30 yards.

The reserve is notable for its butterflies, 23 species have been recorded, including the common blue, marbled white and green hairstreak. This last is our only native butterfly to echo the colour of our landscape! It is a very difficult butterfly to follow, with a quick, erratic flight. The caterpillars feed on dogwood, broom, gorse or sometimes bramble.

Slow-worms are also a feature of this reserve, and many birds are to be heard and seen.

Management
A considerable amount of scrub removal has been undertaken to free the juniper in the hope that it will regenerate. Much more clearance has still to be done and the lines of the old trackways and the parish boundary are to be opened up.

WINDSOR HILL

OS sheet 165; SP 826027

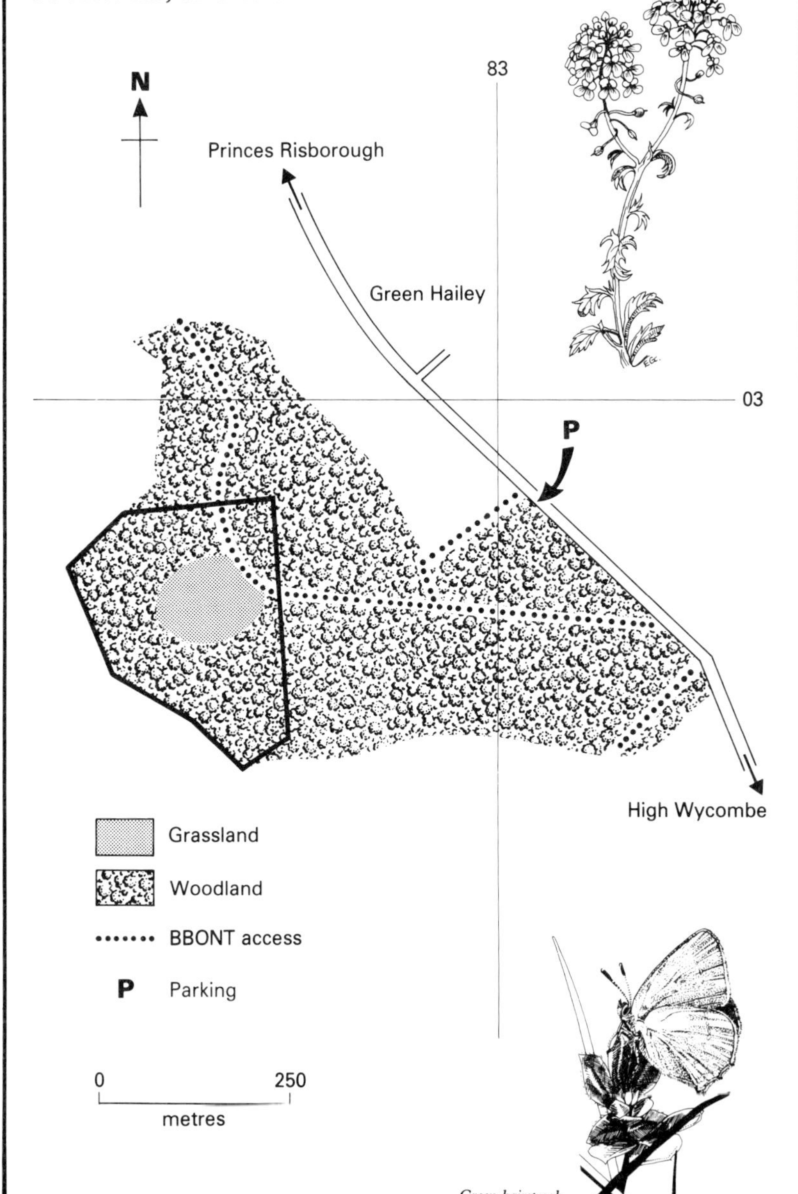

Wild candytuft

Green hairstreak

OXFORDSHIRE Reserves

Asham Meads
Banbury Sewage Farm
Blenheim Farm Community Nature Reserve
Bullingdon Bog (Lye Valley)
Chinnor Hill
Cholsey Marsh
Dry Sandford Pit
Foxholes
Glyme Valley
Harry Williams'/Lamb's Pool Reserve
Hartslock
Henry Stephen/C. S. Lewis Reserve
Hitchcopse Pit
Hook Norton Railway Cutting
Horley
Hornton Meadows
Iffley Meadows
Lashford Lane Fen
Lewknor Copse
Matthew Arnold Field
Oakley Hill
Oxey Mead
The Slade, Bloxham
Sydlings Copse and College Pond
Tuckmill Meadow
Vicarage Pit
The Warburg Reserve
Warren Bank
Westwell Gorse
Whitecross Green Wood
Woodsides Meadow, Wendlebury

ASHAM MEADS

OS sheet 164; SP 593139

Nearest towns Oxford and Bicester

Three old meadows, totalling 22.8 hectares (56 acres) in area, an SSSI bought by BBONT in 1987.

Location
Leaving Oxford on the A421 for Bicester, take the turn for Islip and then proceed through Oddington, Charlton and then Murcott. Continue towards Boarstall, and after about ½ mile turn off south along the track past Whitecross Green Farm. There is a small car park on the reserve.

Access
Open to the public, but please keep to the tracks or field edges to avoid unnecessary trampling. Dogs must be kept on a lead.

Description
The reserve consists of three meadows, Upper Marsh, Lower Marsh and Rowbotham, with the same boundaries as they had on a map of 1590. It is thought that they belonged to the lost Otmoor village of Nash, from which comes the name Asham.

The fields look tawny compared with the more vivid green of the highly fertilised single crop arable fields around them. Their hue results from the large variety of wild grasses and flowers growing.

Their plant communities reflect slight changes in soil conditions. Thus, Upper and Lower Marsh are dominated by crested dog's-tail grass (in flower 6–7) and black (common) knapweed (8–9). Rowbotham however has meadow foxtail grass (5–6) and the attractive great burnet (6–9). Upper Marsh was arable a century ago, whereas the other fields were grassland then and today carry more wildflowers. Rowbotham is the wettest and has several plants such as dyer's greenweed (6–8) and devils'-bit scabious (6–10) not found in the other fields.

The 'ridge and furrow' produced by ploughing in the past helped drainage and also raised the ridges above the seasonal flooding. Recent drainage of neighbouring fields has lowered the water table; even the ponds on the borders of the reserve are now dry for most of the year (but that between Lower and Upper Marsh remains full). However, the furrows are deep enough to be waterlogged each Winter, and they contain a different plant community from the ridges, with tufted hair-grass, jointed rush and other plants tolerant of waterlogging. Rowbotham lacks ridge and furrow but is not level.

The meadows and their fine surrounding hedges are valuable for butterflies and other invertebrates. Snipe and other waders feed here in Winter and in Summer willow and reed warblers nest in the small withy bed. The small oakwood was planted 50 years ago, and the blackthorn thicket nearby has a colony of black hairstreak butterflies.

Management
BBONT is continuing the traditional management of late hay cutting followed by Autumn grazing by cattle.

ASHAM MEADS

OS sheet 164; SP 593139

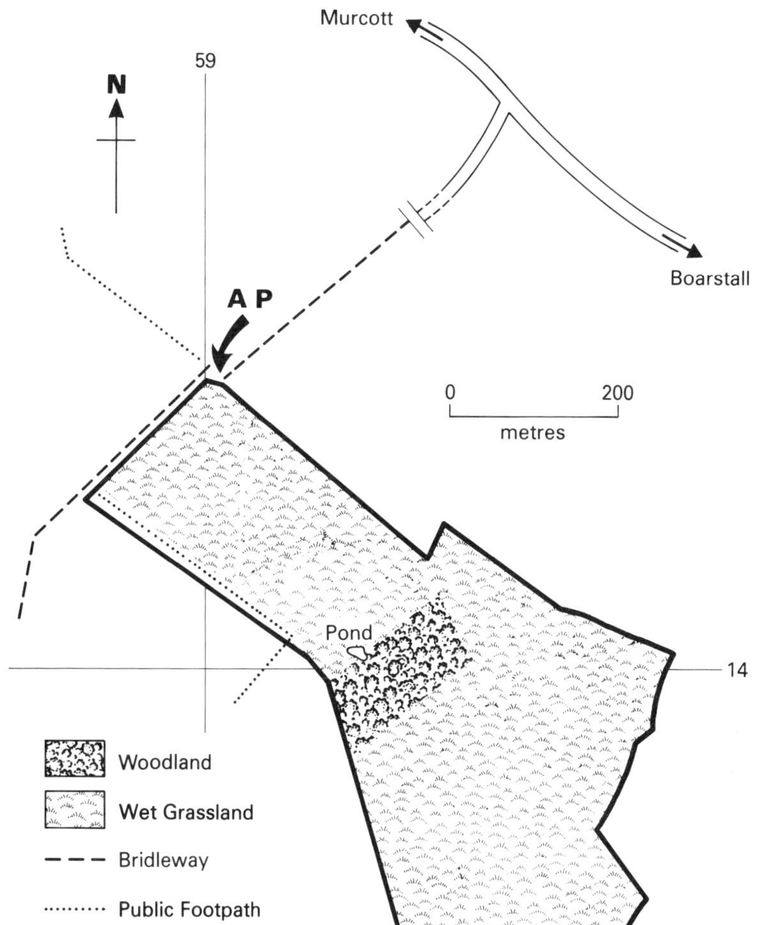

BANBURY SEWAGE FARM

OS sheet 151; SP 474405

Nearest town Banbury

Five large meadows, 27.4 hectares (67.8 acres) in area, managed by BBONT since 1980 by agreement with the Thames Water Authority.

Location
Leave Banbury to the east on the A422 which goes to Brackley and Buckingham, but just past the railway take the right fork to Overthorpe, Warkworth and Kings Sutton. After half a mile there is a track leading off to the right at right angles to the road. This forms the western boundary of the reserve. Parking is as shown on the map.

Access
Open to BBONT members; a public footpath runs down the western boundary and crosses the reserve. The site is inevitably rather smelly.

Description
The reserve consists of five large meadows which are botanically rather disappointing. Some of the meadows have been ploughed and all are regularly flooded as part of their function in connection with the adjoining sewage works. Grasses such as cock's-foot and meadow foxtail dominate, with such attractive wildflowers as cuckooflower (flowering 4–6), meadow buttercup (4–9) and common vetch (5–9). The ditches support colourful freshwater wildflowers including brooklime (5–9) and celery-leaved buttercup (5–9).

At the southern tip of the reserve are two small plantations of alder, wild cherry and field maple. These were planted by BBONT and in time will provide valuable shelter and nesting sites for song birds.

Bird life provides the special interest of the reserve. Up to 300 mallard have been seen at any one time, as have about 100 teal, snipe and lapwing. Detailed observations by members of the Banbury Ornithological Society have resulted in 130 different species being recorded. Look out for dunlin, Bewick's swan and golden plover in Winter, and curlew, common sandpiper and greenshank on migration in Spring and Autumn. The rare lesser yellowlegs was recorded in December 1983, and it would always be worth looking out for other unusual passage migrants.

Lapwing

Management
Meadows not used for sewage treatment are grazed in Summer.

BANBURY SEWAGE FARM

OS sheet 151; SP 474405

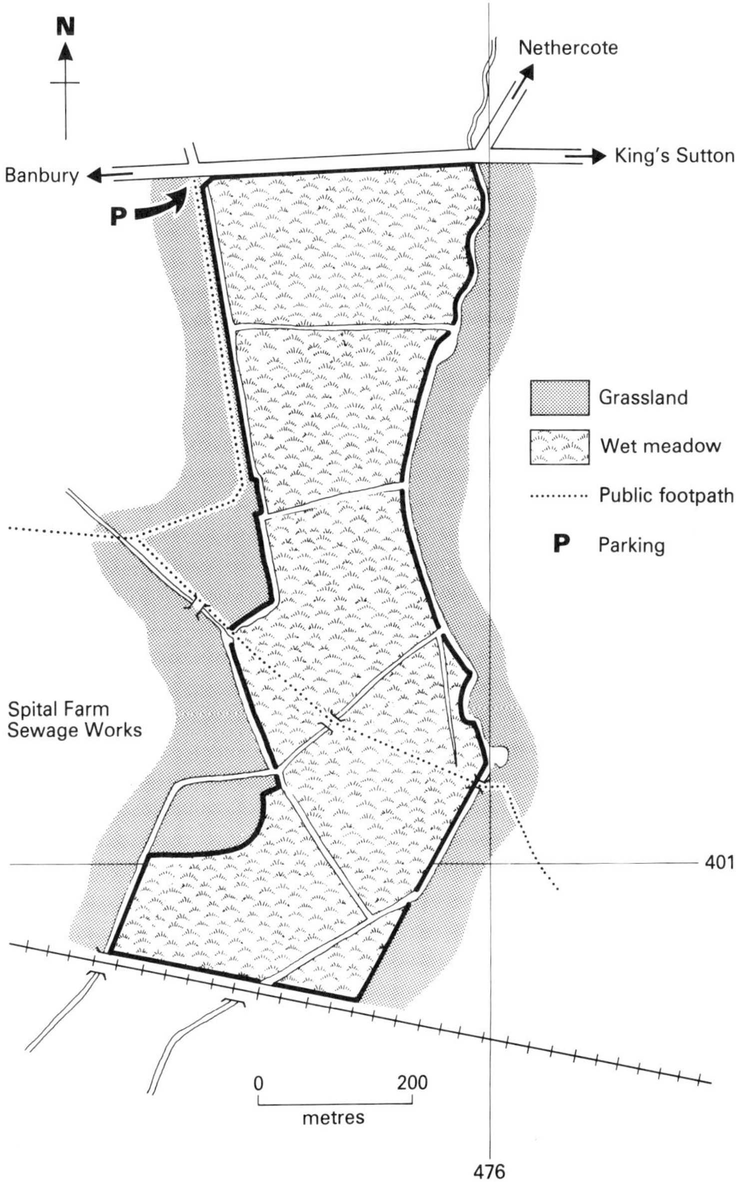

BLENHEIM FARM Community Nature Reserve

OS sheet 164; SP 365195

Nearest town Charlbury

A field of 0.5 hectare (1⅓ acres), which allows BBONT to demonstrate various management techniques. It was given to BBONT in 1987.

Location
Approaching Charlbury from Oxford or Witney, by-pass the centre by following the main road around the east of the town. The road plunges down a steep hill, on the right of which are laybys for parking. Blenheim Farm Cottage is in the bottom of the valley; a public footpath runs beside it up the valley and through the reserve.

Access
Open to the public.

Description
The field lies at the top of a dry valley on Jurassic limestone. It is bordered by hedgerows and a small, scrubby woodland which is good for warblers and other songbirds. Species heard include the blackcap with a rich melodious song, and the chiffchaff. The song of the latter, a repetitive 'tsip tsap tsap tsip', is hardly musical but welcome for all that for this bird is a reliable augur of Spring and usually first heard at the end of March.

The old hedges contain many species of trees and shrubs including field maple and crab apple. The true wild crab apple is often rather thorny, with blossom appearing in May. Its small round apples are yellowish, sometimes flushed with red, and very sour. Dogwood and spindle are also seen in these hedges, reflecting the limey nature of the soil here.

The grassland sheltered by these hedges carries a good show of wildflowers partial to the limey soil, such as cowslips (flowering 4–5) and lady's bedstraw (7–8). 51 grassland plants and 16 woodland plants have been recorded so far.

The owners gave the field to BBONT as they wished to ensure the survival of the area as an 'unimproved' meadow available for local people to use for walking, studying and enjoying. For some time BBONT had been discussing the possibility of setting up some 'Community Nature Reserves' in the three counties – reserves which need not be of great conservation value, but are near a town or village and accessible. These reserves would be run by and for the local community.

This field met all the criteria. Several local people expressed interest in the site and a reserve supporters/management group has now been set up.

Management
The grassland will be divided into plots to demonstrate how different cutting regimes can encourage different species and increase the overall variety. The hedges are leggy and gappy in places and BBONT will be planting native species to thicken them. There is to be some coppicing of the wood and elder will be removed.

BLENHEIM FARM Community Nature Reserve

OS sheet 164; SP 365195

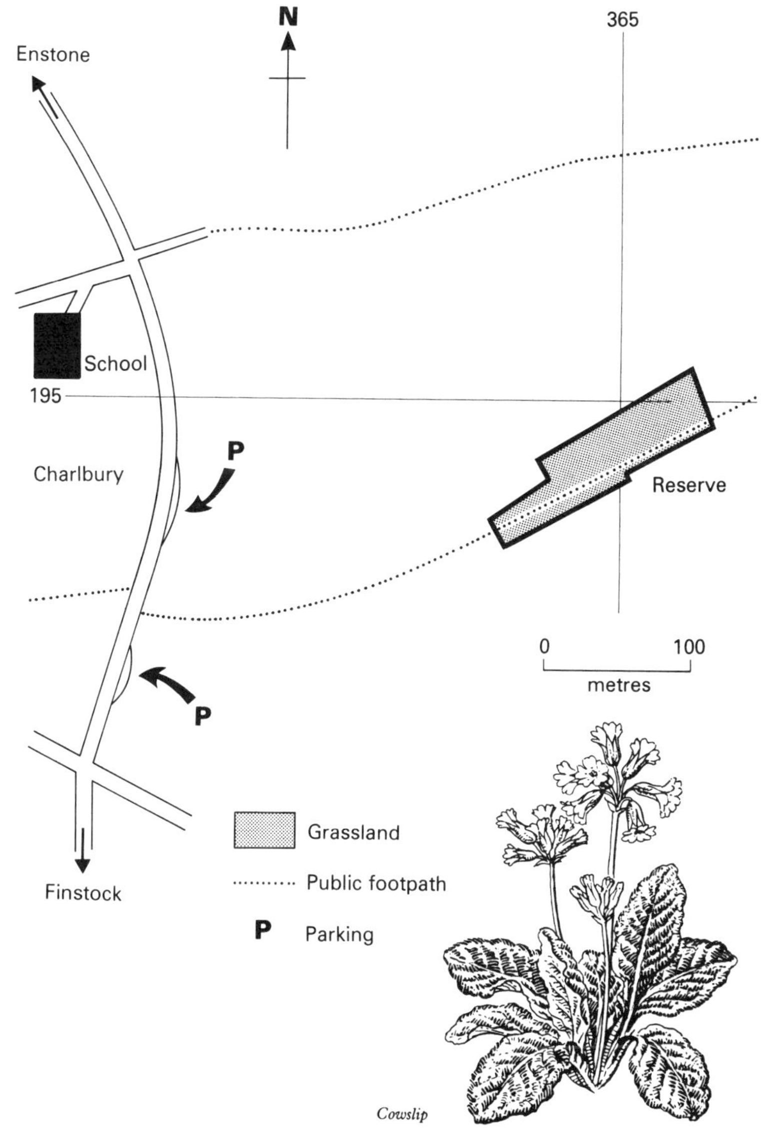

Cowslip

BULLINGDON BOG (Lye Valley)

OS sheet 164; SP 548058

Nearest town Oxford

A patch of fen, about 0.4 hectare (1 acre) in area. It forms part of a larger SSSI. BBONT has leased it from Oxford City Council since 1972.

Location
The reserve is within Oxford's boundaries. Take the road from Oxford City through Headington to the second set of traffic lights. Turn right, cross the roundabout, and take the next turn to the right into Girdlestone Road and park. The reserve lies 300 yards down the valley on your left.

Access
Open to the public.

Description
This is a fine example of a calcareous valley fen. It is a relic of areas of wetland which once fringed lowland rivers and streams but which have gradually been drained over the centuries to create farmland. There are few areas like this still to be found in the Midlands, and even fewer that are within a city's boundaries.

The site is alongside one arm of the Lye Brook, and is fed by water seeping out from rather porous rock beds where they meet impervious clay. Pollen grains preserved in the waterlogged conditions show that long ago the area was open water which became invaded by reeds. As the reeds died down they accumulated under the surface and, because decay was retarded in the airless conditions, peat accumulated and is now 1½ yards deep in places. Where this peat is exposed a 'fen' developed with wet-tolerant plants growing on the waterlogged ground.

In the course of time this vegetation has been succeeded by scrubby wet woodland or carr and the ground has become drier.

Bullingdon Bog has been a popular site since botany became a science in the 17th century. Being so near Oxford it was well known to amateur botanists in Victorian times. As a result, more than 300 plants have been recorded here although not all are still to be seen. The scene is set by sedges and rushes but there are many colourful flowers to interest the visitor, such as marsh helleborine (flowering 7–8) of the orchid family, grass-of-Parnassus (7–10), marsh lousewort (5–9) and the handsome butterwort (5–7). Like the better known sundew, this last plant catches small insects. They are trapped by a sticky secretion exuded by the leaves and their soft parts are digested.

There is much insect life, including some rare beetles and spiders. Reed warblers and reed buntings are among the birds breeding on the reserve.

Management
The site suffers from erosion, fires and fly tipping and occasionally has to be cleaned of rubbish. Encroaching willow scrub also needs attention.

BULLINGDON BOG (Lye Valley)

OS sheet 164; SP 548058

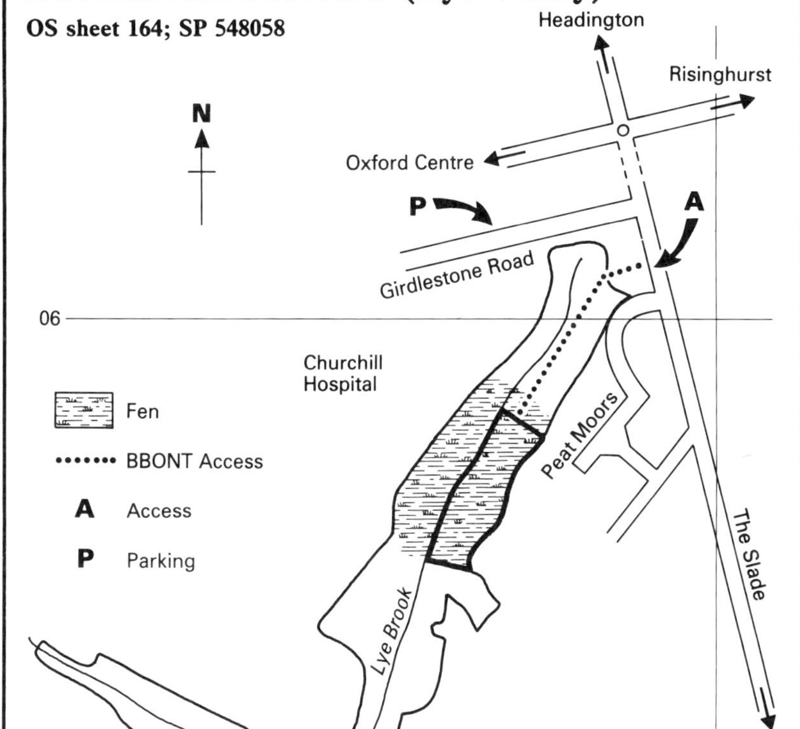

Marsh helleborine

CHINNOR HILL

OS sheet 165; SP 766005

Nearest towns Chinnor, Princes Risborough and Thame

A reserve of 28 hectares (70 acres) of chalk slope, with woodland, scrub and grassland. It is an SSSI, owned by BBONT since 1965. In addition to its wildlife interest there are two Saxon burial mounds and three ancient sunken tracks down the slopes.

Location
From Chinnor take the unclassified road up the scarp and take the first left turn back towards Bledlow Ridge. As the road goes round to the right, turn left again into Hill Top Lane and drive to the car park at the end. The reserve is in front of you.

Access
Open to the public. There is a network of public footpaths and bridleways, including a length of the Upper Icknield Way.

Description
The reserve is situated on the Chiltern escarpment and carries soils of two kinds, clay-with-flints at the top and shallow, quickly draining chalky soil on the slopes. There is an area of woodland with oak and ash, and small areas of mature open-floored beech woodland at the top of the hill. These apart, the whole area was once open grazing land, with juniper growing on the steep slopes.

Earlier this century, livestock grazing ceased and, in the 1950's, the rabbit population was decimated by myxomatosis. With this grazing pressure removed, a dense mixed scrub of hawthorn, wild privet, traveller's-joy, spindle, buckthorn and other woody species has invaded large areas and has killed some juniper, although much survives.

There are wild roses in the grassland lying to the south of the Icknield Way; dog rose (flowering 6–7), sweet briar (6–7) with apple-scented leaves and burnet rose (5–7) can be seen. This last has purple-black hips.

The open chalk grassland carries typical wildflowers, such as wild thyme (flowering 5–8), common rock-rose (5–9) and wild candytuft (7–8). Pyramidal (6–8), frog (6–8) and bee (6–7) are amongst the orchids seen. Look also for the straw-coloured flower of the very prickly carline thistle (7–10).

As is to be expected from such a large area of scrub, the reserve has good populations of nesting birds, including seven different species of warblers. In Autumn and Winter the areas of scrub provide a rich larder of berries for birds of the thrush family including redwings and fieldfares.

With its fine views this reserve is worth a visit at any time of year.

Management
Some scrub has been cleared, some of the hazel has been coppiced and the grassland area at the summit is grazed each year by BBONT's sheep. The intention is to clear more of the scrub. This is one of the reserves recommended for school use.

CHINNOR HILL

OS sheet 165; SP 766005

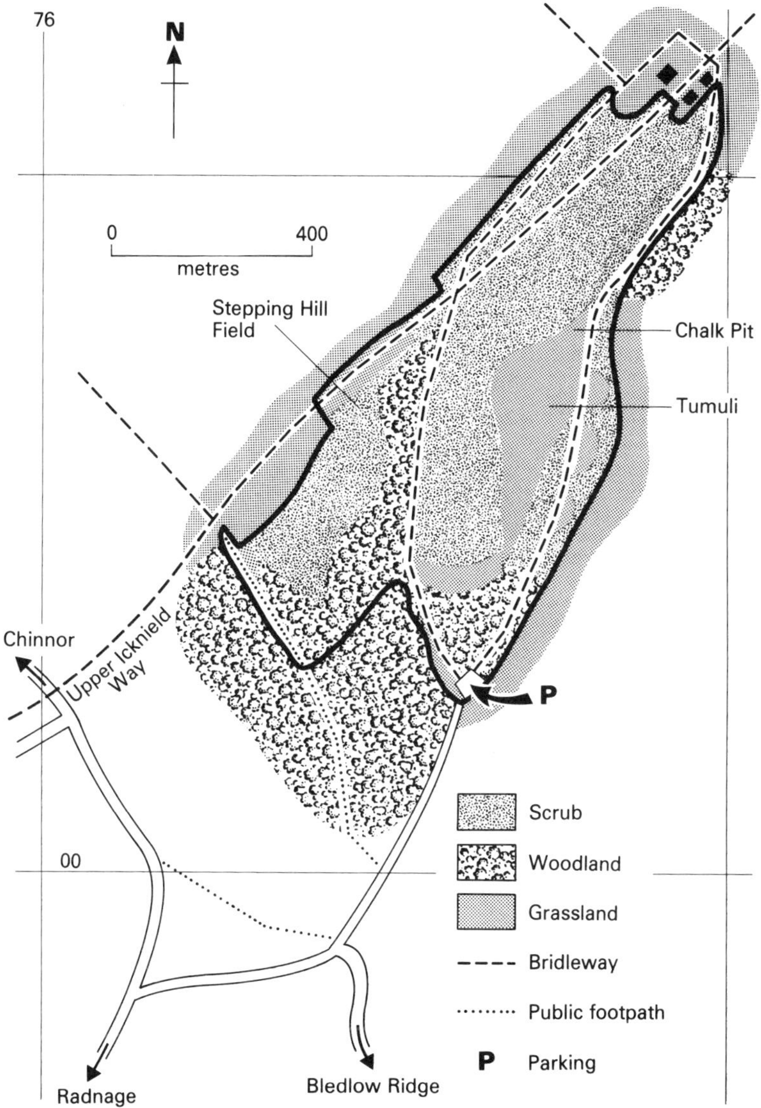

CHOLSEY MARSH

OS sheet 174/175; SU 604857

Nearest town Wallingford

An area of marsh, 11.9 hectares (29.4 acres) in area. It has been leased to BBONT since 1986.

Location
Leave Wallingford southwards on the A329. Just past the Fairmile Hospital at Cholsey turn left and drive down to the River Thames and park on the verge. The main part of the reserve is on your left; there is a long tail running downstream to the railway viaduct at SU 595848.

Access
Open to the public. Dogs should be kept on leads.

Description
At first sight a rather featureless area, this reserve is in fact full of interest. It is a relic of a type of habitat which was common along our river systems but which has now been largely drained. Several rather different areas can be identified within it – one which has little else but pond sedge growing, for example, another of 'tall fen' which is marked by meadowsweet (flowering 6–9). The tassels of reed sweet-grass can be seen in places (6–8).

The birds to be seen provide great interest, for the reserve supplies the requirements of several different species. Snipe breed here, and their unusual 'drumming' display flight can be watched in Spring and early Summer. Reed buntings have been recorded as nesting.

Sedge warblers also breed here in fair numbers. They are Summer visitors of course. When preparing to migrate in Autumn, they will be seen feeding busily on the insect life of the reserve, laying up fat for the journey, the success of which will depend on it. A sedge warbler's normal weight is just over third of an ounce. Even if fat reserves increased this by 50% or more, it will be over a tenth of an ounce lighter by the time it reaches North Africa, 18 hours and 1000 miles after departure. It will now have to stop and feed before facing the 30 hour crossing of the Sahara Desert. If its weight drops to a third of an ounce, its metabolism is strained and it is likely to die.

Swallows also congregate to roost here in late Summer. Their sudden departure (which amazed people in the days before the secrets of migration were known) is a sign of impending Autumn.

It is from late Summer through to Spring that the site really comes to life, attracting a good many birds apart from the swallows. Cholsey Marsh is a favoured Winter roosting site for wagtails and reed buntings. Well over a hundred meadow pipits a time are often seen and there are also large numbers of corn buntings regularly arriving to roost – the Marsh is a Winter stronghold for this bird in this part of Oxfordshire.

Management
Rubbish tipping has been countered.

CHOLSEY MARSH

OS sheet 174/175; SU 604857

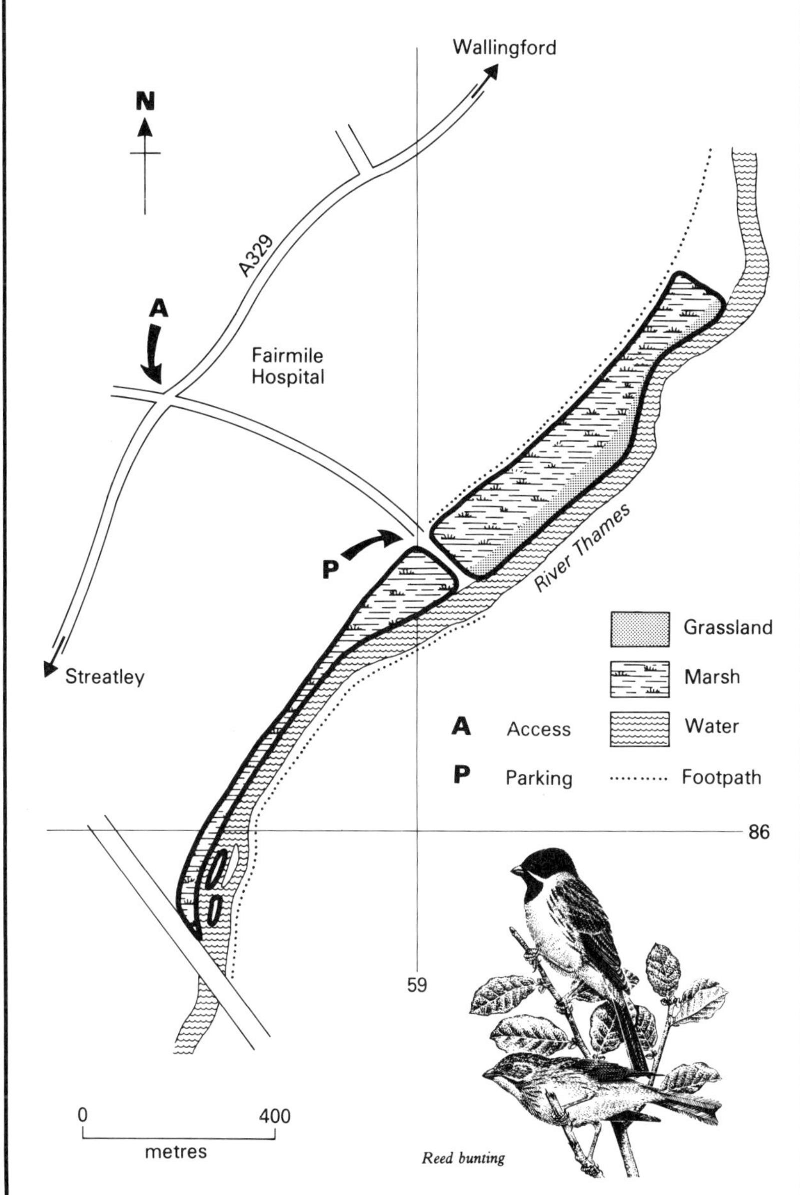

Reed bunting

DRY SANDFORD PIT

OS sheet 164; SU 467995

Nearest towns Oxford and Abingdon

An old stone and sand quarry, the northern part of which is an SSSI as a result of both its ecological and its geological interest. 8 hectares (20 acres) in area, BBONT bought the freehold in 1979.

Location
The reserve lies in Cothill village, 2 miles north-west of Abingdon. It is best approached by turning off Wootton Road (B4017) almost directly opposite the garage which marks its junction with the Boars Hill/Foxcombe Hill road. Very soon after entering Cothill, turn left opposite a curio shop into a layby which leads up a slope to a small car park. The reserve is in front of you over a stile.

Access
Open to the public, but the fen area is a fragile habitat and should be avoided. Keep dogs under control and on a lead when BBONT's sheep are grazing. **WARNING**: keep clear of the cliff overhangs which may suddenly give way!

Description
Largely an old sand and stone quarry which has been partly filled in; low cliffs still remain on the northern side. Please do not attempt to collect fossils from the cliff face. These exposed areas are one of Britain's best sites for burrowing bees and wasps.

Springs feed part of the old quarry floor and there is a small pond of shallow open water (in which can be seen abundant growths of the stonewort *Chara*) surrounded by areas of fen. The stream flowing from these Springs joins Sandford Brook which forms the western boundary of the reserve, the part adjacent being mainly woodland and scrub. Apart from two small ponds, the rest is grassland with a light scrub cover in places.

The fen area and its surrounds have been partly colonised by various species of willow, birch and other trees, and even some Scots pine. The botanist will find this an interesting area, with early marsh-orchids (flowering 5–7) and common spotted-orchids (5–7) and a large colony of marsh helleborines (7–8). These wet areas provide a habitat for a wide range of wildlife, including frogs, toads, newts, grass snakes and water voles, while birds include moorhen and kingfisher (seen quite frequently). Dragonflies and damselflies are common.

The turf areas carry many wildflowers. On the bare rocks and sandy soil below the cliff a 'heath' of lichens can be seen with many fungi in the Autumn.

Glow-worms are seen on Summer nights, and the penetrating song of the great green bush-cricket is something to listen for, from July to October. The reserve also holds a large colony of marbled white butterflies.

Management
Scrub growth around the pond and fen areas is checked while the BBONT sheep graze the grassy areas and limit the spread of coarse herbage.

DRY SANDFORD PIT

OS sheet 164; SU 467995

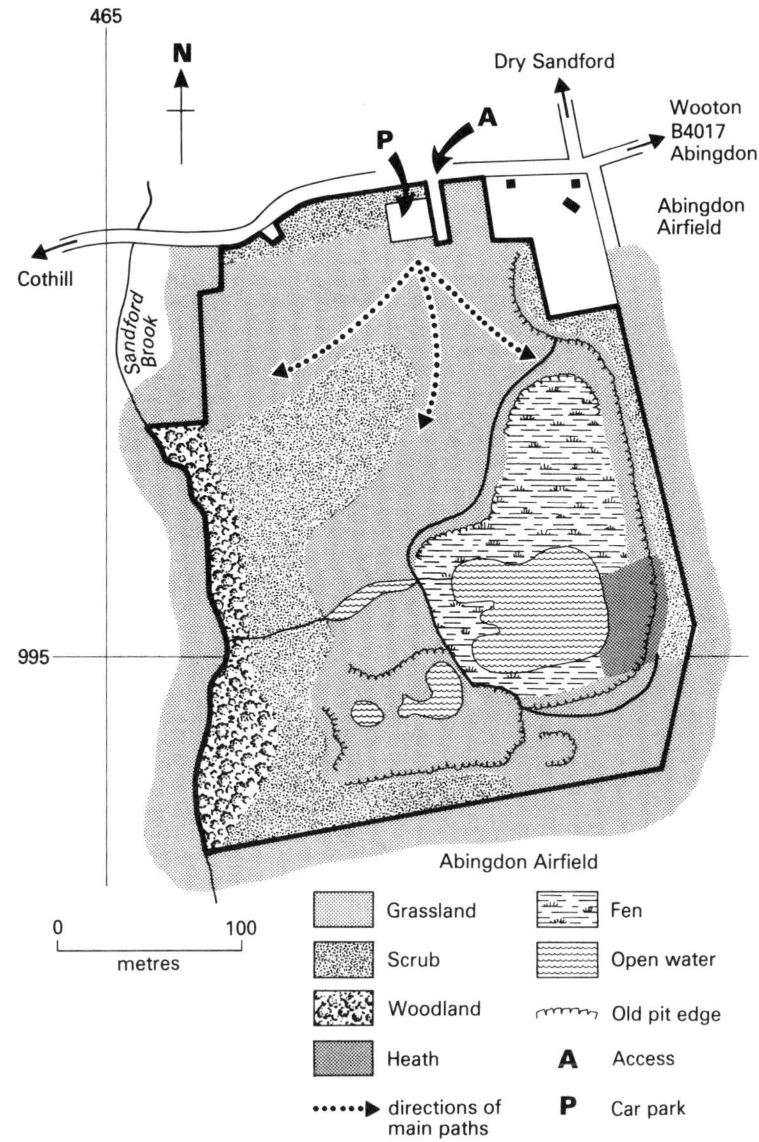

FOXHOLES

OS sheet 163; SP 254206

Nearest towns Burford and Chipping Norton

A reserve of 64 hectares (158 acres) encompassing a variety of habitats and a wet meadow. Much of the reserve is of SSSI status. Part of the reserve is owned by BBONT and part was leased in 1975 on a 99 year lease.

Location
From Burford take the A424 towards Stow-on-the-Wold. Take the third lane on the right and just before Bruern Abbey turn left along a track following the western edge of Cocksmoor Wood. After half a mile, the car park is reached, on the right at SP 259208.

Access
Open to BBONT members on the waymarked route. It is open to the public only on the public footpaths and bridleways.

Description
Although Bould Wood proper is not actually part of the reserve, local people know the whole woodland area, including Foxholes reserve, as 'Bould Wood'.

The woodland is a remnant of the ancient Wychwood Forest, sloping gently from higher ground with acid soil across base-rich (alkaline) soils to the River Evenlode. Soil and management over the centuries have combined to create a variety of habitats. In the part of the reserve next to Cocksmoor Wood, for example, two kinds of wet ash-maple woodland can be recognised. Both hold other trees and shrubs in the coppice to confuse the visitor, but the clue lies with the wildflowers. In one, primrose flowers (3–6) together with early-purple orchid (4–6), common twayblade (6–7), and wild angelica (6–9), while the other (where oak and ash standard trees are seen) carries plants not common elsewhere in the wood, including sanicle (5–8), goldilocks buttercup (4–6) and herb-Paris (5–7). It provides a good example of habitat difference.

Elsewhere in the reserve there are a few mature oaks, but much has been planted with beech, hornbeam and larch. The Bruern estate (from which BBONT leases) has thinned the beechwood area in Ladies Ride and replanted with mixed, mainly deciduous, trees.

The woodland is very good for fungi. In the open rides on the acid soils plants such as purple moor-grass, bog moss, and heath spotted-orchid (flowering 6–8) can be seen. Uncommon mosses and liverworts are also found along the rides. In the wet meadow alongside the river, great burnet (5–8) and devils'-bit scabious (6–10) flower.

Over 20 species of butterfly have been recorded, including the wood white. Muntjac, fallow deer and foxes use the reserve, and both the grass snake and adder are common. The bird life is abundant and varied.

Management
The rides must be maintained and horse barriers repaired. Some old ash coppice is being re-coppiced. Nest boxes are maintained. The meadow is grazed and the newly planted willow is protected.

FOXHOLES

OS sheet 163; SP 254206

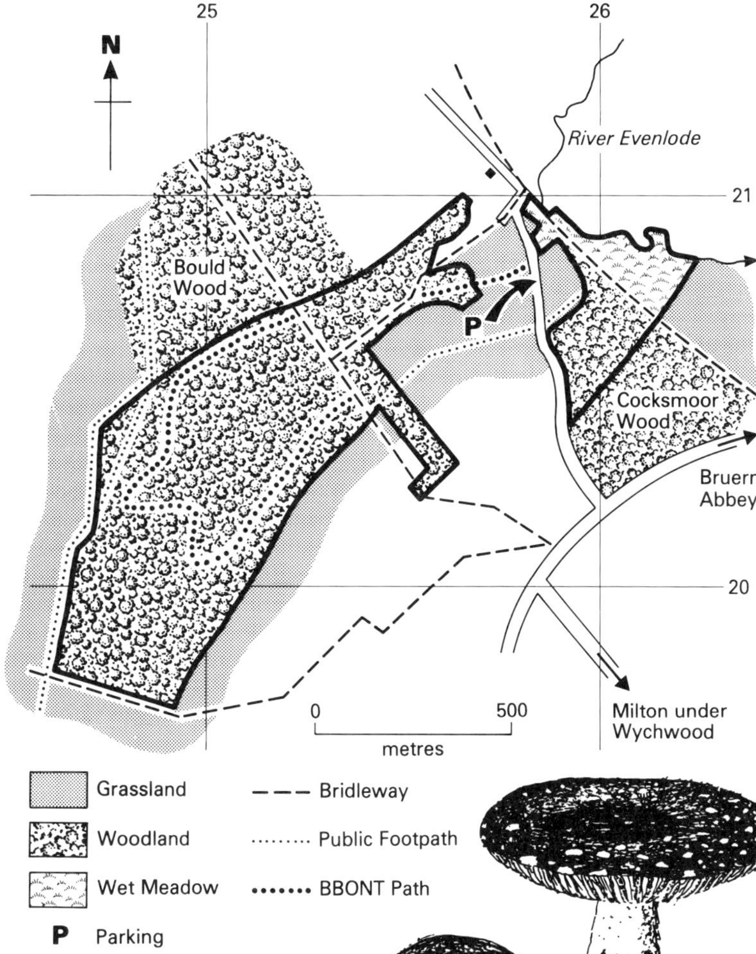

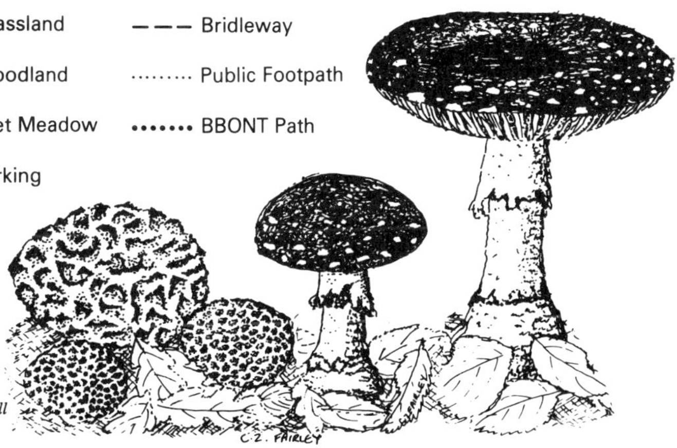

Common earth-ball and fly agaric

GLYME VALLEY

OS sheet 164; SP 335260

Nearest town Chipping Norton

The BBONT reserve, leased from Thames Water in 1988, forms part of an SSSI. 2.3 hectares (5.7 acres) of secluded valley, part of the headwater reaches of the River Glyme.

Location
The A44 leaves the north end of Chipping Norton; after about 1¼ miles it joins the A34 between Stratford and Oxford. Before this, however, about ¾ of a mile past the hospital at the edge of Chipping Norton there is a layby (before the golf course). Park here and walk down the public footpath that leaves here and passes New Chalford Farm on the way to the reserve.

Access
Open to the public, but keep dogs under control if sheep are grazing.

Description
A remote meadow in north Oxfordshire, Glyme Valley consists of limestone grassland with a patch of dense scrub and the River Glyme itself. Few patches of such grassland remain on the edge of the Oxfordshire Cotswolds – a great pity because they are extremely beautiful and rich in wildlife.

Look out for dwarf thistle (flowering 7–9), woolly thistle (7–9), oxeye daisy (5–9), cowslip (4–5), great burnet (6–9) and perforate St. John's-wort (7–9).

At the time of going to press, this site has only been a reserve for a few months and detailed surveys remain to be undertaken. So, any interesting sightings would be gratefully received by the BBONT Office.

Cowslip

Management
Management of this reserve has not been worked out at the time of going to press.

GLYME VALLEY

OS sheet 164; SP 335260

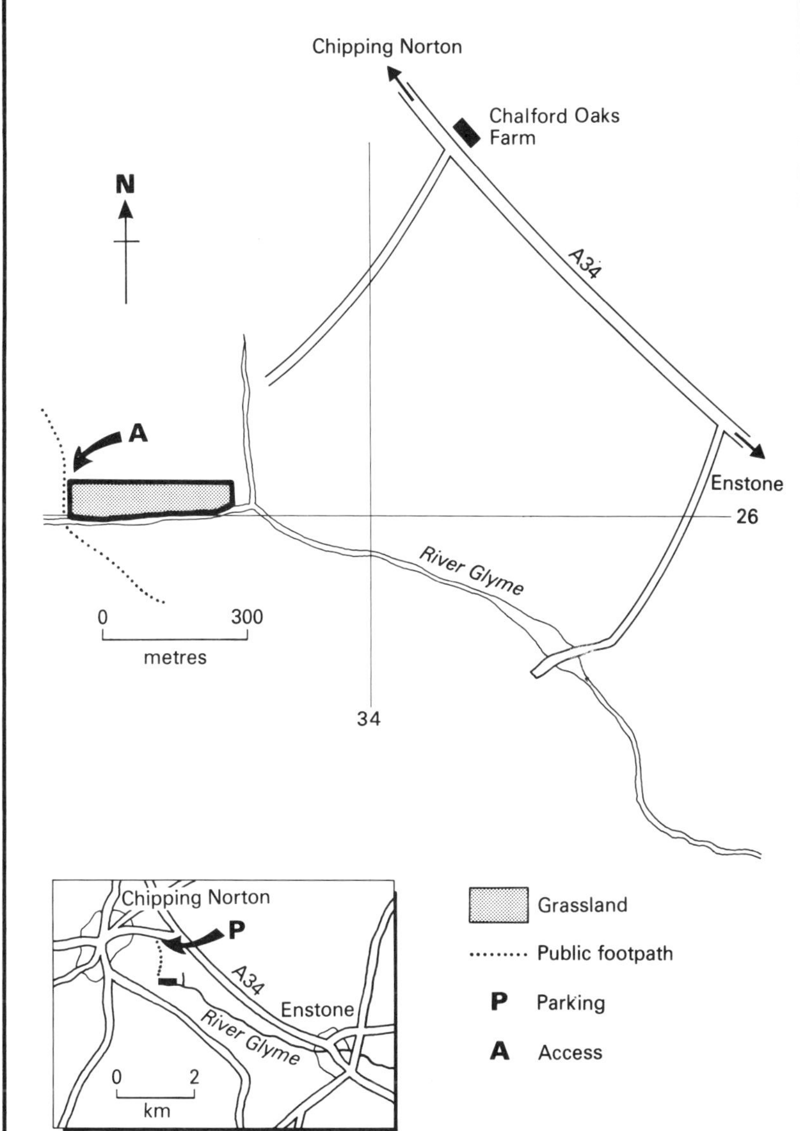

HARRY WILLIAMS'/LAMB'S POOL RESERVE

OS sheet 151; SP 353362

Nearest towns Chipping Norton and Banbury

A reserve of 2 hectares (5 acres). It was given to BBONT by Mr and Mrs J. A. Lamb of Sibford Ferris. It is also named in memory of Harry Williams who was voluntary reserve manager until his death in 1987, and who was a well-known and respected local naturalist.

Perhaps this is the place to pay our respects and give our thanks to those who lend BBONT their skills and time to help safeguard the reserves for the benefit of all of us.

Location
You need to locate Hook Norton, a village which lies to the north of the A361 between Chipping Norton and Banbury (but closer to Chipping Norton). Take the unclassified road running north from Hook Norton towards Sibford Ferris; go over the cross roads past the pub called 'The Gate Hangs High', continue to the bottom of the hill and park on the left verge. Cross the stile and continue on foot along the left-hand hedge to where there is a gate into the reserve.

Access
Open to the public, but visitors are asked to respect plant and bird life and not to use the north bank during the nesting season from 1st February to 31st July. Please keep dogs on a lead.

Description
This is a man-made reservoir with a dam along the south side and western end, fed by water taken from the stream which forms the head of the River Stour. The rest of the reserve consists of a narrow strip of land around the reservoir.

Aquatic plants have been introduced and have successfully colonised the shallow (eastern) end of the reservoir. Here bur-reed now grows, easily recognised by its round spiky fruiting heads. Handsome yellow iris (flowering 5–7) and kingcup or marsh-marigold (3–5) are welcome wildflowers. Like many wetland species they are no longer so easy to find growing in the general countryside outside nature reserves. The greater reedmace also grows here, recognised by its brown sausage-like heads. Most people know it as the 'bulrush' but botanically another less striking plant species owns that name. The confusion arose with the famous painting of 'Moses amongst the Bulrushes' by the Victorian artist Alma-Tadema. To add visual excitement, he painted this reedmace!

This reserve is now a fine place for birds – coot, moorhen, dabchick, mallard and tufted duck, mute swan and Canada goose have all nested around the open water and on the island. Nests of 19 other species have been recorded elsewhere on the reserve.

Management
The bank along the south side needs periodic attention to repair the dilapidation caused by wind and wave action. Willows are pollarded and the easten bank is mown to prevent colonisation by scrub.

HARRY WILLIAMS'/LAMB'S POOL RESERVE

OS sheet 151; SP 353362

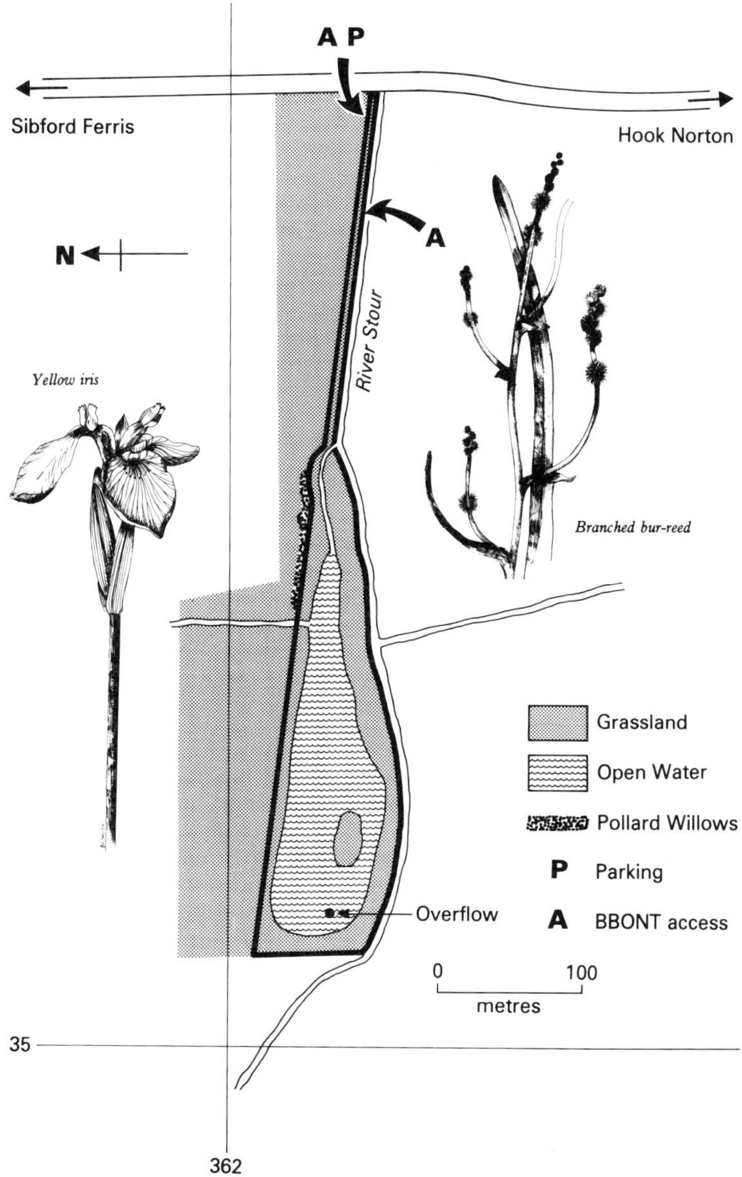

HARTSLOCK

OS sheet 175; SU 618795

Nearest towns Goring and Streatley

An area of rich chalk grassland, 4.4 hectares (11 acres) in area. It is of SSSI status (part of a larger area of SSSI) bought by BBONT in 1975.

Location
You need to find Goring Railway Station. If approaching from Oxford this means taking the A329 to Streatley. Turn left at the traffic lights, cross the river into Goring and straight on past the shops until a T-junction, where you turn right. The station is ahead.

Drive past the station for a mile and park off the road by the first or second junction. Walk up the right hand track to a gate on the right that leads to the reserve; this track must be kept clear as it is used by farm vehicles. On a sunny day it is simpler to walk from Goring, if you like the exercise.

Access
Open to the public. Please keep dogs on a lead at all times when the BBONT sheep are grazing. Please do not cross the temporary fencelines which mark areas where research into such things as the effects of grazing is being carried out.

Description
This reserve lies on steep slopes overlooking the Goring Gap, where the Thames cuts through the chalk downs.

From afar, the reserve is the characteristic tawny colour of old, traditionally grazed grassland, neither ploughed nor artificially fertilised. Its hue, which can be seen for example from the Oxford–London trains which cross the Thames on Brunel's bridge not far away, comes from the rich variety of many wild grasses and plants. Common rock-rose (flowering5–9), harebell (7–9), bird's-foot trefoil (6–9), and the uncommon bastard toadflax (6–8) grow here; the thin red stems of dodder (7–9) (it is parasitic on clovers) is amongst those to be seen.

Indeed, the tally includes an endangered species. A temporary warden, employed by BBONT with financial aid from sponsors mounts 24-hour guard while the plant is in flower.

This is what modern conservation must (sadly) sometimes be about.

Management
When BBONT acquired the site, scrub had begun to colonise.

Although scrub can provide birds with secure nest sites, it does mean the loss of light-loving grassland flowers and consequently the butterflies and other insects reliant on them.

The quality of the grassland is now maintained by sheep grazing. Fencelines are erected to contain their grazing, and the effect of grazing at different times of year is monitored by the success or otherwise of the wildflowers and wildlife in the various compartments.

HARTSLOCK

OS sheet 175; SU 618795

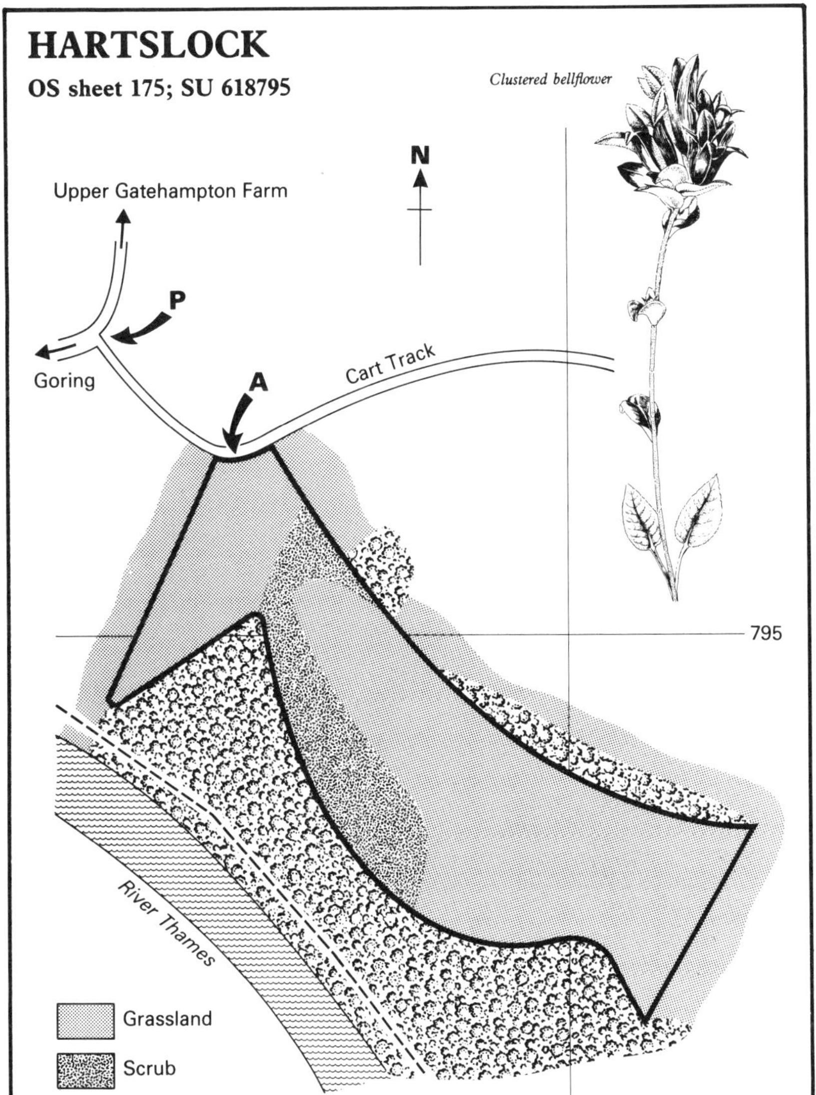

HENRY STEPHEN/C. S. LEWIS RESERVE

OS sheet 164; SP 560065

Nearest town Oxford

A large man-made pond with mixed woodland surrounds, 3 hectares (7½ acres) in area. The freehold was bought by BBONT in 1969. The reserve is named after C. S. Lewis, an Oxford figure of some renown (and author of the classic children's book 'The Lion, the Witch and the Wardrobe' amongst others) and Henry Stephen, a professor of the University, who lived in Lewis Close. There is a field study hut for school parties.

Location
From the Headington roundabout on the A40 north-east of Oxford go south on the ring road, take the first turn on the left and then go immediately right into Kiln Lane. After about 300 yards turn right into Lewis Close. Park at the end, and the reserve is at the end of the path.

Access
Open to the public.

Description
This ground was once owned by C. S. Lewis, a friend of J. R. R. Tolkein. It is said that the worlds of Narnia and Middle Earth (Tolkein's famous Hobbit territory) were dreamed up here. So the site remains as a memorial to imagination as well as being of great wildlife interest.

There is certainly plenty of natural history to be seen here. When entering the reserve, you are faced with a large pond created during the last century when clay was dug. This pond is deceptively deep and under no circumstances should visitors wade into it. It is extremely busy with aquatic plants. The spectacular displays of damselflies and dragonflies alone are worth a long journey in Summer months. Apart from these insects, the pond is interesting for the moorhens and little grebes (dabchicks) which regularly nest here.

A small stream flows in from the east to add aquatic variety, and giant horsetail grows at its silty margins.

On the slopes above, beech, oak, birch, alder, sycamore and larch can be seen. Here numerous woodland birds are to be heard in Spring and early Summer, and the scuttlings of small mammals at any time of year.

And look out for the 'doggers'. These are large, rounded, sandstone concretions littered on the slopes amidst the trees.

The reserve is used regularly by school parties for a variety of environmental projects and a WATCH group meets here once a month.

Management
The invasive sycamores are removed and 'habitat piles' are created from felled trees and brushwood – we hope the rotting wood will attract a host of beetles and other invertebrate life. The pond will need to be dredged in part, and the overhanging trees will occasionally need to be cut back to reduce shading. Larches at the top of the reserve are being removed and young oak, rowan and a few beech trees are being planted.

HENRY STEPHEN/C. S. LEWIS RESERVE

OS sheet 164; SP 560065

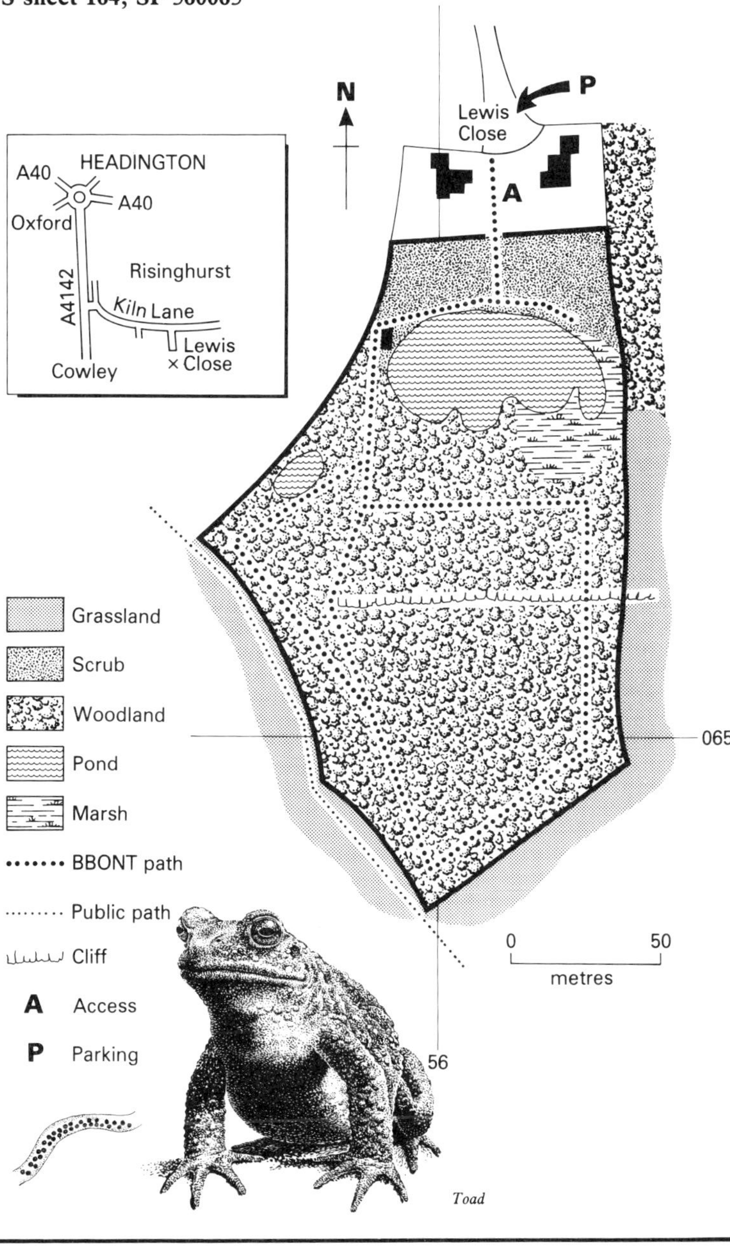

Toad

HITCHCOPSE PIT

OS sheet 164; SU 452996

Nearest towns Abingdon and Oxford

A disused sandpit. The reserve covers 0.8 hectares (2 acres) and has been managed by BBONT by agreement with the owners since 1979.

Location
From Oxford, take the A420 towards Faringdon. Fork left on the A338 towards Wantage. After a quarter of a mile park on the right hand verge and take the bridleway on the other, left hand side of the road. The reserve is 400 yards down this bridleway, lying to your left behind a thick hedge.

Access
Open to the public. Dogs must be kept on a lead. Beware of the danger of rockfall around the lips of the pit.

Description
This long disused sandpit is full of wildlife interest. It is close to Frilford Heath, a famous entomological site with exceptional numbers of ant, bee and wasp species. It complements DRY SANDFORD PIT reserve as it satisfies species which require more open, fine sand.

The holes of burrowing wasps can be seen honeycombing the cliffs and other exposed surfaces. The holes are often clustered in 'colonies' although the wasps are themselves solitary. The nest does not have a hierarchy of queen and workers, instead each mother collects her own food and the visitor may sometimes be able to watch a wasp dragging a caterpillar or other prey back to her nest to provide a larder for the grubs that will hatch from her eggs. These burrowing wasps are more slender than the familiar common wasp, and not all carry yellow and black stripes.

Solitary mining bees may also be seen, many of which look like small honey bees. It must be remembered that insects emerge fully grown from the pupal stage – a small fly or bee is never a youngster but of a different species.

There are also some rare and localised beetles. Although few would wish to handle a live beetle, they do attract their own devotees. At the turn of the century some clerics were quizzing an eminent biologist on his view of God. 'He has an inordinate fondness for beetles' came his considered reply. Indeed, there are thousands of different species of beetle to be found in Britain, in almost every habitat.

There are many attractive wildflowers including viper's-bugloss (flowering 6–9). Evening-primrose (6–9) can also be seen – its large flowers open at dusk to be visited by night-flying moths which pollinate them.

There is a shallow pond which supports the great crested newt.

Management
Many problems have arisen in the past by illegal motor bike scrambling and, to a lesser extent, by horse riding. Both activities have been reduced by the erection of a fence.

Some scrub has been cleared from the bottom of the pit.

HITCHCOPSE PIT

OS sheet 164; SU 452996

Evening-primrose

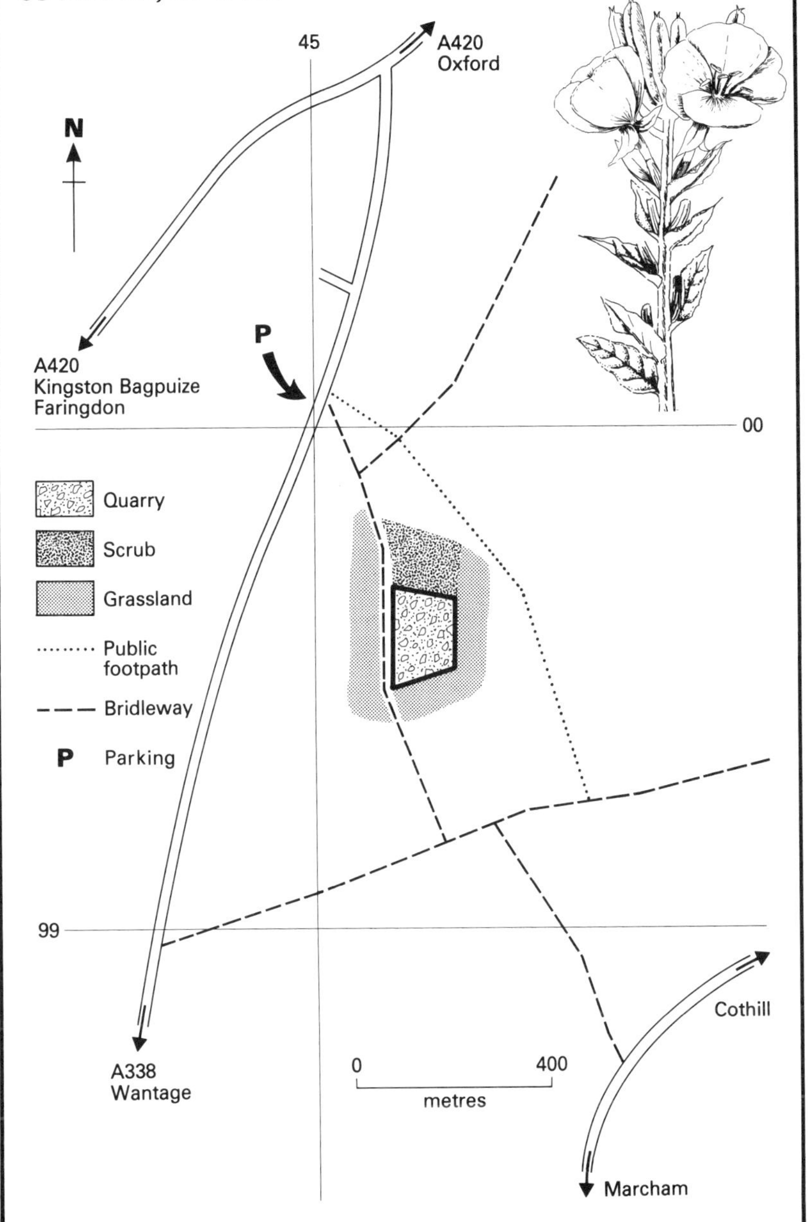

HOOK NORTON RAILWAY CUTTING

OS sheet 151; SP 360320

Nearest towns Banbury and Chipping Norton

A length of disused railway line, totalling 7.7 hectares (19 acres). All but the part north of the road bridge at SP 360323 is an SSSI, scheduled mainly for its geological interest. The site was bought by BBONT in 1972.

Location
From Banbury take the A361 towards Chipping Norton. About 2 miles after South Newington turn right along the unclassified road through Swerford and towards Hook Norton. After 1 mile, fork right. Cross the railway bridge and park in the layby on the right. Access to the reserve is on the other side of the bridge.

To reach the southern section, continue in the direction of Hook Norton, then turn sharp left; at the T-junction turn right, and then left after 50 yards. Park down the hill by the remains of the railway bridge. Access is by the stile on the left.

Access
Open to the public. The railway line was used to transport iron ore from the Hook Norton area to blast furnaces in the Midlands and South Wales; it was closed in 1963 after quarrying ceased. The tunnel separating the two sections does not belong to BBONT and is out of bounds.

Description
The geological interest lies in the exposures of Jurassic oolitic limestones, which have been used in house and stone walls in the area. The red staining is due to the presence of iron oxide.

Three main areas can be recognised. North of the road bridge, on the embankment, is a developing woodland of oak and maple to the west and scrub and woodland to the east. The centre section is also developing woodland, here dominated by ash and willows. Male fern and hart's-tongue fern are to be seen here.

In Spring and Summer these areas are alive with bird song. 47 species have been recorded, and amongst those nesting are woodpeckers (great spotted and green), garden warbler, blackcap, whitethroat and goldcrest. For the specialist, the retaining walls along the track support many lichens.

The southern section provides what is essentially limestone grassland with scrub encroaching. Amongst the many wildflowers kidney vetch (flowering 6–9) and wild carrot (6–8) are abundant.

Butterflies are numerous including marbled whites and the cutting is noted for its populations of bees.

Management
On the northern sections the vegetation on each side of the railway track is cut in alternate years. Much of the invasive scrub has been removed from the southern section, which is grazed by sheep.

HOOK NORTON RAILWAY CUTTING

OS sheet 151; SP 360320

Green woodpecker

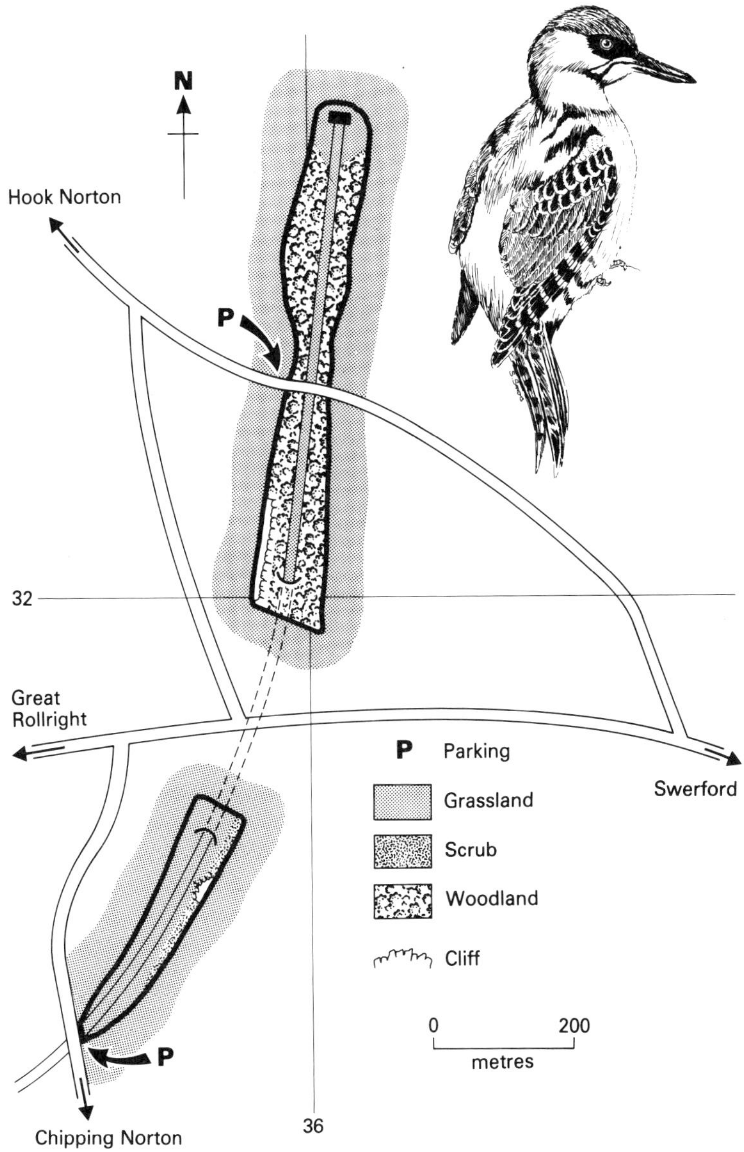

HORLEY

OS sheet 151; SP 412430

Nearest town Banbury

An old railway line, siding and quarry workings, 12 hectares (30 acres) in area. BBONT has leased it since 1982.

Location
From Banbury take the A41 north and fork left on the outskirts along the A422. Continue to Wroxton where you take the third turn on the right on the western outskirts along an unclassified road. Parking is as shown on the map.

Access
Open to the public.

Description
The reserve consists of an old railway line and siding with old ironstone quarry workings alongside. When the railway line was abandoned, the natural process of succession began and scrub started to invade. At the same time, some areas of the site were used for dumping.

There are a variety of habitats. Areas of dense scrub are beginning to develop into woodland on some of the steep banks. Scrub species such as hawthorn and blackthorn (sloe) have seeds which are easily spread by birds. They grow (and sloe suckers) to form a dense cover into which trees seed, equally randomly. Shaded as they are, the growth of the tree saplings may be slow at first but the scrub can protect them from rabbits and other grazing animals. After about 60 years from bare ground, tall woodland is in place.

The reserve also carries two larch plantations. Larch foliage lets quite a bit of light through and larch woods often contain other woody species. Here we have hazel, elder and holly as well as some grey alder.

There are areas of rough derelict land where the dumping has taken place. Although no very unusual plants grow here, cover is provided for mice, voles and shrews as well as for some birds.

The short grassy bank and track and the flat grassy area carry many wildflowers.

In all, 170 different plant species have been recorded at Horley, 43 species of breeding birds, and 11 of butterflies, including the marbled white.

Management
Some scrub control is undertaken to maintain the areas of open grassland. Ideally this should now be grazed, preferably by sheep. From a conservation point of view sheep are easier to manage (especially on a mixed site such as this) and rather more beneficial than cattle. With the correct stocking rates, sheep can produce a fine sward which encourages the growth of the shorter wildflowers. Cattle tend to tear wholesale at the plants but avoid cow pats (the sites of which become marked by thistles) and their weight 'poaches' wet ground, turning it into a muddy morass.

This reserve is one of those recommended for school visits.

HORLEY

OS sheet 151; SP 412430

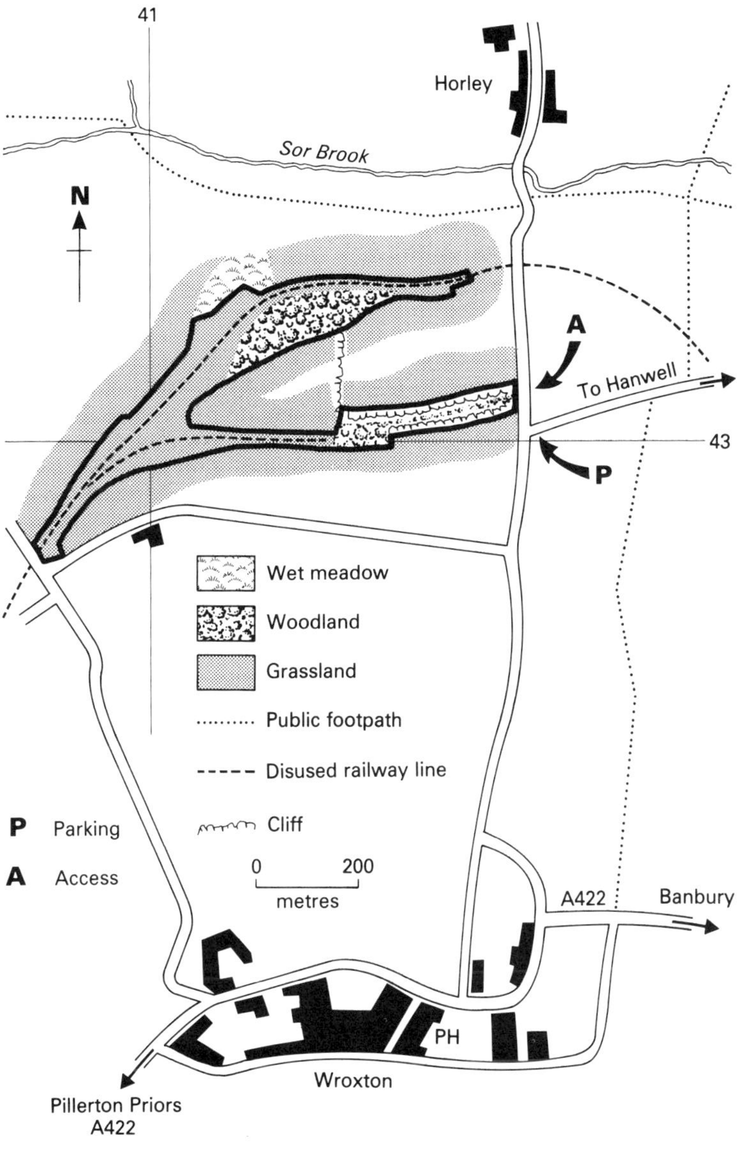

HORNTON MEADOWS

OS sheet 151; SP 397466

Nearest town Banbury

A pair of ancient meadows and an ancient hedgerow covering 6 hectares (15 acres) in all; bought by BBONT in 1988.

Location
Follow the A422 out of Banbury towards Wroxton. At Wroxton take the second right towards Horley. Go straight through Horley following a sharp bend to the left near the church and travel about a mile and a quarter. On the right is a bridleway just before a set of crossroads. Walk or drive down the track as shown on the map, and then walk along the boundary of the field to the reserve. The bridleway is unsuitable for cars after heavy rain and turning can be difficult.

Access
Open to the public.

Description
The reserve lies in a valley and consists of two damp hay meadows surrounded by old hedgerows which contain a variety of different species of woody shrubs.

The number of these can very often be used to give us an idea of the age of the hedge. It is marked into 30 yard lengths and a count is made of the number of different species in each length. Count all wild roses as one and ignore climbers. The average number per 30 yards is the age of the hedge in centuries. Hedgerow history can be complicated however, and not all hedges lend themselves to this handy method – those which were by origin strips of woodland left when fields were cleared at each side for example. See also the remarks under OAKLEY PARISH HEDGE.

The two meadows are flower-rich. There is an abundance of great burnet (flowering 6–9) whose distinctive dull crimson cylindrical flower heads make a good 'indicator' of old unimproved meadows of this kind, and pignut (flowering 5–6), its underground tubers are the 'nuts in May' of the familiar children's rhyme.

It is interesting that large numbers of frogs are to be found here in the Spring. Our countryside was formerly dotted with ponds and pools. However, field ponds are usually filled when agricultural 'improvement' takes place and this, together with the general drainage of marshy areas, has caused even the 'common' frog to become less than common in today's countryside. We can no longer take frog spawn for granted.

The reserve is also notable for dragonflies and damselflies. Birds recorded as nesting in the hedges include the tree pipit.

Management
The hay is cut in traditional fashion.

HORNTON MEADOWS

OS sheet 151; SP 397466

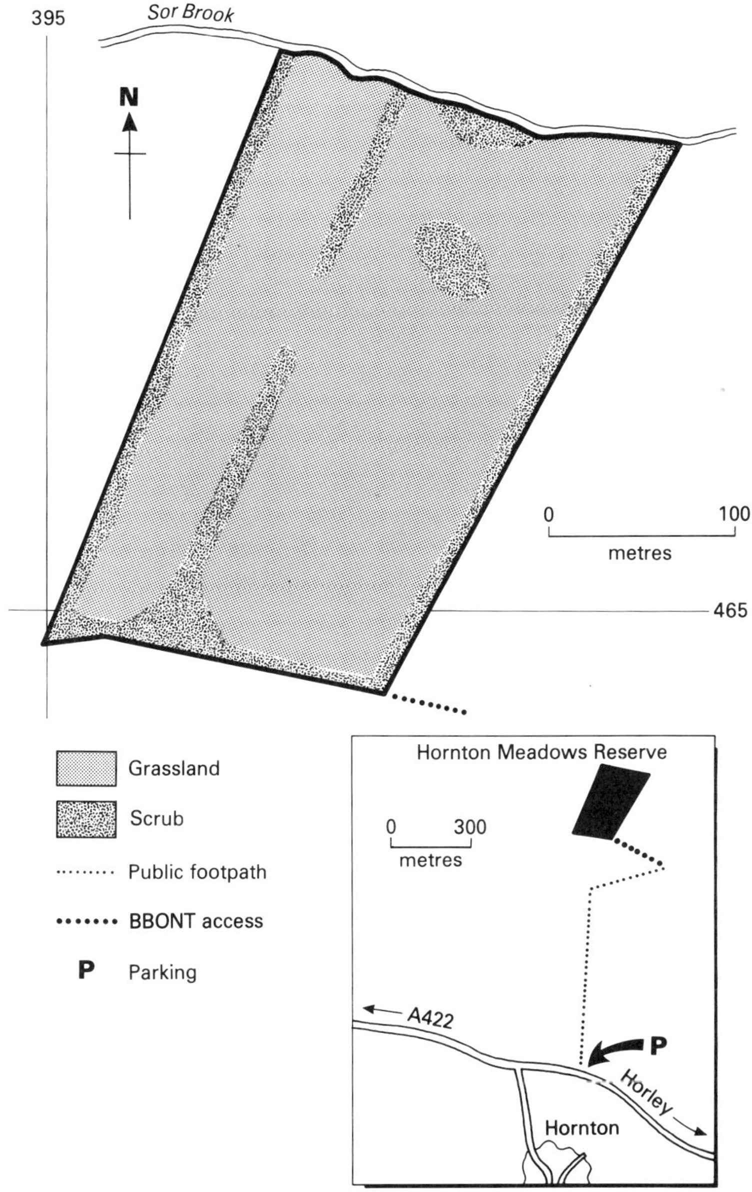

IFFLEY MEADOWS

OS sheet 164; SP 524037

Nearest town Oxford

Ancient wet meadowland, 33 hectare (82 acres), famous for its fritillary flowers. An SSSI, it has been leased from Oxford City Council since 1983.

Location

The reserve may be reached by parking within Iffley village and crossing the Thames via the bridge and lock gates at Iffley Lock. If so, it is worth making a detour to see the Norman carvings of Iffley Church.

Alternatively, before reaching Iffley village from central Oxford turn right at traffic lights into Donnington Bridge Road. Park off this road in Meadow Lane before the bridge. Walk across the bridge and along the towpath. The reserve extends to the ring road. It also includes the triangular field to the south of the ring road.

Access

Open to the public. Please take care not to trample the fritillary plants, especially when they are in flower. Dogs must be kept on a lead.

Description

The main interest of these ancient wet meadows, embraced by two arms of the River Thames is the presence of the fritillary, in flower in late Spring (4–5). It used to flower profusely across the meadows, but is now seen mainly around the centre of the section north of the ring road. Numbers were dropping before BBONT assumed control, not only as a result of picking but also from lack of appropriate management of the meadows. These meadows are one of the few remaining places in Britain where this plant can be seen flowering in its natural surroundings (it is possible to buy reared bulbs in garden centres for garden planting).

These atmospheric meadows are crossed by old river channels and willow-lined ditches. Much of the area has clay soil which is still enriched each year by silt brought by flooding. Fritillaries apart, they carry a wide variety of plants many of which are 'indicators' of old meadowland undamaged by farming improvements. These include adder's-tongue, great burnet (flowering 6–9), common meadow-rue (7–8), pepper-saxifrage (6–8), and creeping-Jenny (6–8).

Parts of the meadow are covered by deposits of gravel which adds a rather acid bite to the soil; here fewer wildflowers are seen. However, plants such as meadow buttercup (flowering 5–8), ragged-robin (5–6) and oxeye daisy (5–9) are fairly abundant everywhere. There is also marsh-marigold to be seen in Spring (3–5).

The field south of the ring road is wetter and marshier than the rest of the reserve and here meadowsweet can be seen flowering (6–9), along with abundant soft rush.

Management

The meadows are mown in July after the seed from the fritillary plants has ripened. The hay is baled and removed and livestock are put on to graze the aftermath until November.

IFFLEY MEADOWS

OS sheet 164; SP 524037

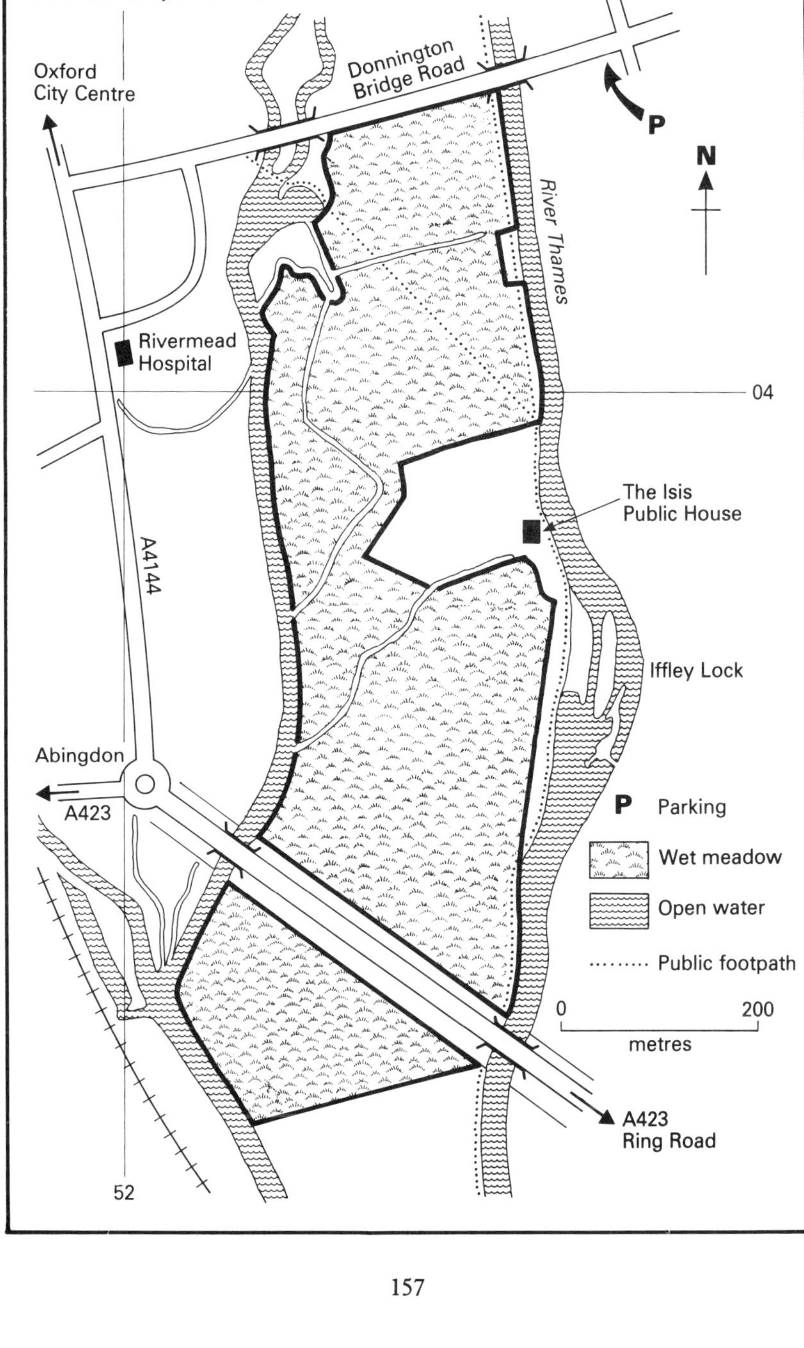

LASHFORD LANE FEN

OS sheet 164; SP 468013

Nearest towns Oxford and Abingdon

A 7 hectare (17.4 acre) wetland SSSI, bought by BBONT in two lots, in 1984 and 1985.

Location
Leave Oxford westwards on the A420 (Botley Road). Go 1½ miles past Cumnor and take the first turning to the left on the unclassified road leading to Abingdon. Take the first turning to the right into Lashford Lane and the reserve is on your right. Parking is allowed at the scout hut about 100 yards further along Lashford Lane on the right hand side.

Access
Open to the public. Dogs must be kept on a lead.

Description
This is a small wetland site in a valley cut down into limestone. The interest comes from the seepage of limey water which has created soil conditions suitable for a plant-rich valley fen. The fen is one of a mosaic of different habitats, others being woodlands of ash and sycamore, and willow and birch, some hawthorn and other scrub, and areas of flower-rich grassland. Sandford Brook runs through the site.

Rich fen such as we have here – nationally an exceptionally scarce habitat – is to be prized. The plants which set the scene are tall reed and lushly growing great willowherb (flowering 7–8), meadowsweet (6–9) and hemp-agrimony (7–9). Amongst these grow slighter plants, many carrying rather pretty flowers – marsh lousewort (flowering 5–9), butterwort (5–7), creeping-Jenny (6–8) and fragrant orchid (6–7). In the centre of the fen is a rather more acid area which is marked by purple moor-grass.

Great crested newts are present in the centre of the fen. They are now nationally rare, so much so that they are one of a number of protected species that it is illegal to kill or disturb. Today's countryside is generally too hostile for these newts and they rely on refuges such as this nature reserve. Protected as they are, in the open countryside they may still be harmed by 'lawful activities' – which include agricultural practice. This one example is enough to show just how important nature reserves are, and how vital the work of BBONT is, if we want to protect wildlife.

There are three areas of limestone grassland on the reserve where visitors may see many wildflowers, such as dwarf thistle (flowering 6–9), fairy flax (6–9) and burnet-saxifrage (7–8).

The scrub and reeds attract many breeding warblers and other songbirds while the marbled white butterfly is one of many insects and other invertebrates for which the reserve is renowned.

Management
Scrub clearance is needed and the grassland areas need mowing and/or grazing to keep them open. Trees in the fen areas need to be removed and the reed beds cut.

LASHFORD LANE FEN

OS sheet 164; SP 468013

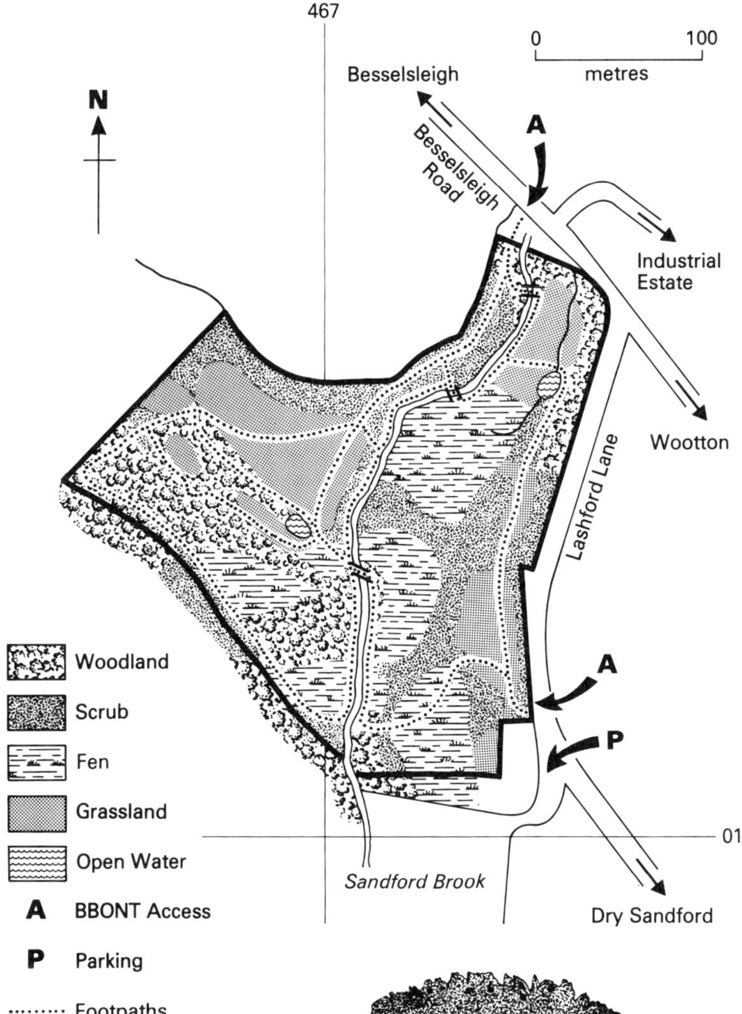

Great crested newt

LEWKNOR COPSE

OS sheet 165; SU 724976

Nearest towns Chinnor, Thame and High Wycombe

A small beechwood of 1 hectare (2½ acres), managed by BBONT by agreement with the owners, dated 1969. It is part of Aston Rowant SSSI.

Location
Detailed directions start from Exit 6 (near Lewknor village) on the M40 between Oxford and High Wycombe. From this exit go east along the B4009 towards Chinnor. After a short distance turn right onto the A40 and after ¼ mile turn right onto the Icknield Way and park on its verge. The reserve is ¼ mile from here, the Aston Rowant National Nature Reserve being on your left as you walk along. It too is well worth a visit. It is easily reached by continuing up the hill on the A40, turning right at the brow, and then right again after about ½ mile (before the cutting leading to the bridge over the motorway). There is a convenient car park at the end.

Access
Open to the public on foot only.

Description
Beechwoods tend to be rather bare beneath the trees because of the dense leaf cover during the Summer, but evergreens can be expected. One species to look for in Lewknor Copse is spurge-laurel. The wood holds a fine colony of this uncommon shrub. Growing up to a yard high, the narrow, leathery leaves are clustered at the tips of the shoots, with the greenish flowers amongst them early in the year (it flowers 1–3). Look also for several species of wild orchids which are typical of beechwoods. The white helleborine (flowering 6–7) whose flowers, held on a characteristically angled stem, barely open. It may be self-pollinating, an advantage where the deep shade reduces the number of insects. Narrow-lipped helleborine (6–7) and common twayblade (6–7) also grow here. About 12 to 18 inches high, they are both inconspicuous with their green flowers, so search carefully.

Bird life in Lewknor Copse is spasmodic. Mistle thrushes sing early in the year, warblers feed in the Spring and various members of the tit family, especially long-tailed tits, visit throughout the year. Finches take the berries of the spurge-laurel (even before they ripen) and the beechmast. In Autumn the distinctive call of the nuthatch may be heard in the canopy of the trees and in a good whitebeam-berry year redwings arrive. Ivy berries provide food for many species at the end of the Winter.

Management
The owners, All Soul's College, Oxford, have thinned the beech trees and felled invasive sycamores. Wild privet (this shrub has leaves which are rather thinner than the oval leaves of garden privet), numerous sycamore seedlings and some bramble have been hand-pulled.

As in many beech woodlands along the Chilterns, some trees are sickly and careful felling and replanting will need to be undertaken.

LEWKNOR COPSE

OS sheet 165; SU 724976

White helleborine

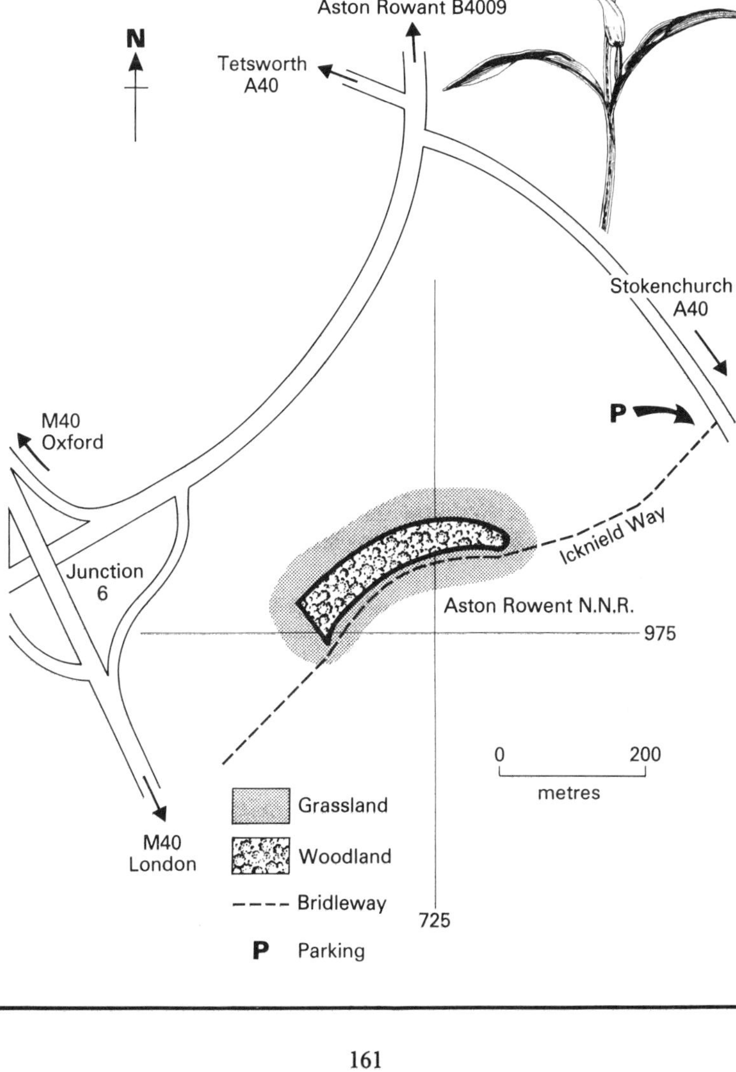

MATTHEW ARNOLD FIELD

OS sheet 164; SP 483023

Nearest towns Oxford and Abingdon

A hollow, now largely invaded by scrub, 1.5 hectares (3.85 acres) in area. It has been managed by BBONT by agreement with the Oxford Preservation Trust since 1981.

Location
Follow the Oxford ring road to the A34 roundabout south-west of the city, then drive west up the hill. Take the next right hand turn, then right again and fork left; keep straight on where the road bends sharply down to the left down Old Boars Hill. Park at the end by Jarn Mound. The reserve is in the fenced hollow on the other side of the field.

Access
Open to the public.

Description
The reserve is named after Matthew Arnold as it is an area described in his poem 'The Scholar Gipsy'. It is said that the land here was once used as allotments by prisoners from the Napoleonic Wars, by whom an (as yet) undiscovered well was dug.

The reserve is in a small hollow where a Spring rises, where a bed of porous sand rests on clay. This Spring feeds a small stream which runs out into the adjoining field. The site, for a long time open grassland, has been colonised by scrub and trees including oak, ash, willow, hawthorn and blackthorn. Acid-tolerant gorse, broom, and wood sedge are also present.

Common spotted-orchid flowers here (6–8) and grass snakes are often seen.

It is an educational nature reserve, primarily for Matthew Arnold School in Cumnor. The pupils are surveying the vegetation types and monitoring the changes which occur as the vegetation on the site develops.

Management
The reserve is managed by teachers and pupils from Matthew Arnold School. They have excavated a small pond in the stream bed, created a marshy area and cleared glades – light reaches the pond to the benefit of invertebrates, frogs and newts. Toad spawn, too, might eventually be seen. While frogs will use any water for breeding, even mere puddles or marshy ground, if a pond is not available, toads seek deeper, more reliable pools.

Additional note
Some neighbouring areas are of interest. One is Jarn Mound and the field above the reserve. Jarn Mound is not as ancient as its name might suggest; it was in fact built in the 1930's to give a view of Oxford.

The Jarn Wild Garden, where many interesting plants may be seen, is owned by the Oxford Preservation Trust. It is on the right at the last junction in the location instructions above.

MATTHEW ARNOLD FIELD

OS sheet 164; SP 483023

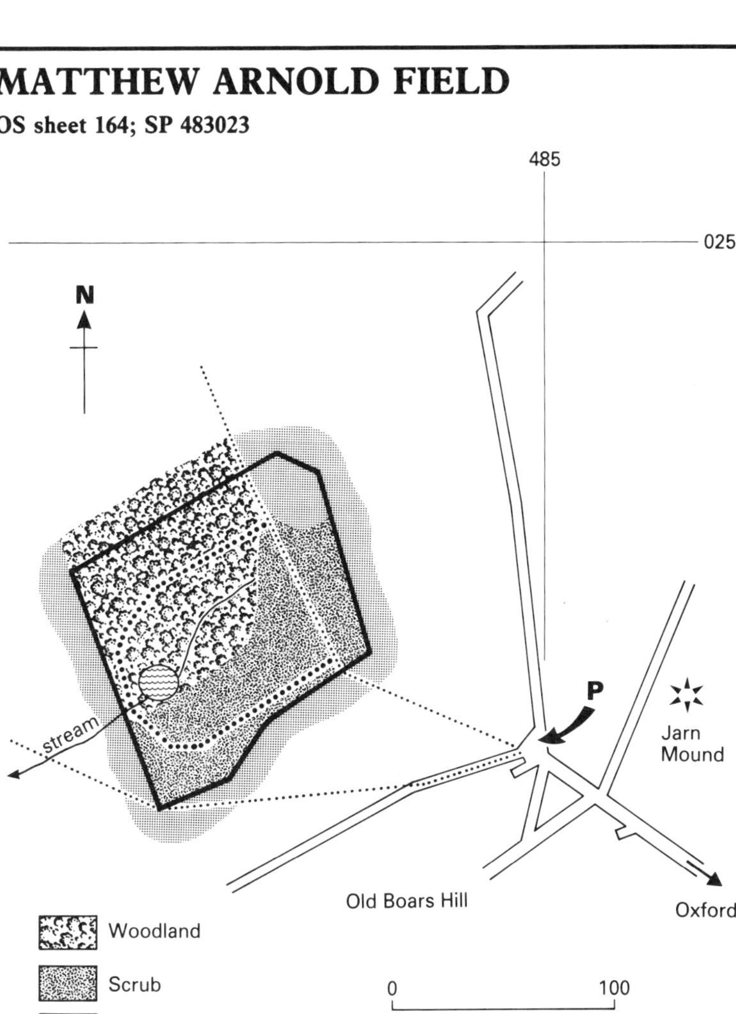

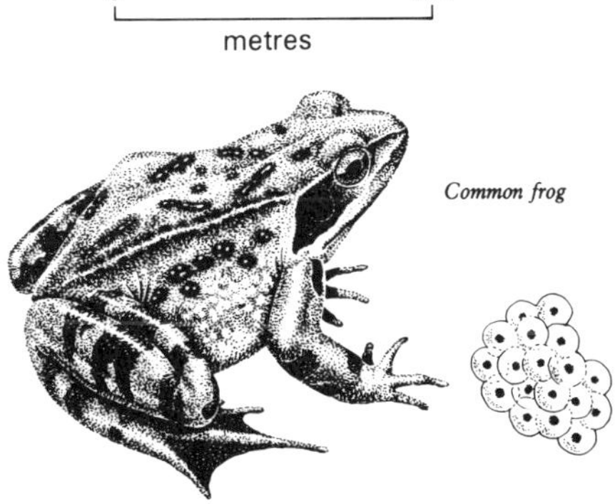

Common frog

OAKLEY HILL

OS sheet 165; SU 755995

Nearest towns Chinnor, Thame and Princes Risborough

A 4.5 hectare (11 acre) area of old chalk downland purchased by BBONT in 1983.

Location
From Thame travel south-eastwards along the B4445 straight through Chinnor to the Upper Icknield Way at SP 760002. Park on the Icknield Way but do not drive along it. Walk south-westwards for 1 mile and turn left round the end of the quarry workings and walk back ½ mile. The reserve is on your right.

Access
Open to the public, but no dogs please for there may be sheep grazing.

Description
This is mainly old chalk downland which has been partially colonised by scrub. There are still open areas with typical downland plants including common rock-rose (flowering 5–9), wild thyme (5–8), Chiltern gentian (8–9), clustered bellflower (7–8), harebell (7–9), common spotted-orchid (6–8), and yellow-wort (6–10). Many butterflies can be expected on the wing in the Summer months.

There is a fringe of beechwood along the top of the slope.

Management
The reserve has suffered neglect for many years, but some of the scrub has been removed and the area is grazed by BBONT sheep. These are the famous sponsored sheep that have appeared in local newspapers, on radio, and even on national TV.

The media (and the public) find the sheep sponsoring scheme intriguing. Any member of the public can sponsor a sheep for BBONT, that is help to pay for its purchase, management and care. Sheep need regular attention to prevent diseases, parasites and other ailments which afflict them. The sponsors can choose their own name for their sheep and even meet it on one of the annual sheep days.

The 150 strong flock of speckle-faced Beulahs, a mid-Wales breed, is supervised by a full-time BBONT Grasslands Officer helped by volunteers. Managing their grazing pattern to benefit the wildflowers and the butterflies and other insects of the grassland reserves requires careful planning. The wildlife interest of the reserves has to be balanced against the health and welfare of the sheep.

Each year BBONT has to decide which reserves are to be set aside, to be allowed to rest ungrazed and come into full flower. During Spring and early Summer, the sheep have to be grazed on other reserves. However, after the wildflowers have set seed the sheep can also be grazed on the reserves set aside, and the reserve managers welcome them – for by then the grass has grown tall.

OAKLEY HILL

OS sheet 165; SU 755995

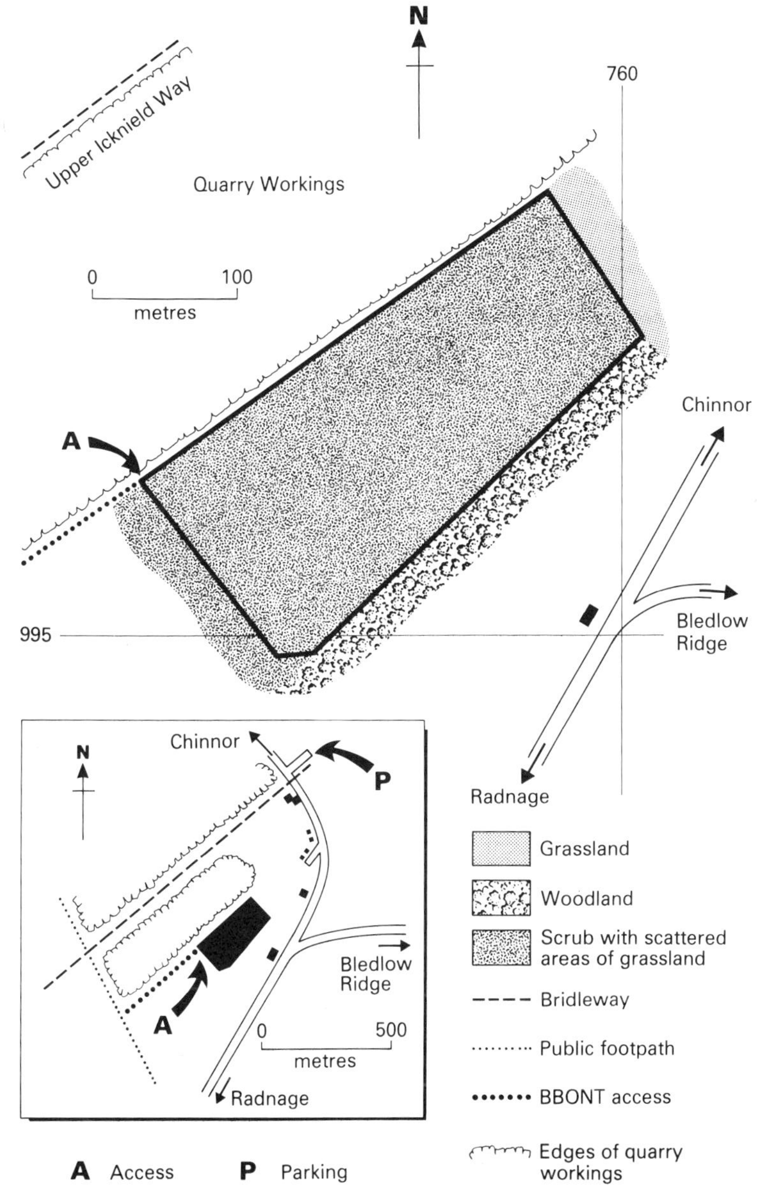

OXEY MEAD

OS sheet 164; SP 480107

Nearest town Oxford

Around 8 hectares (20 acres) of ancient meadowland. An SSSI, it was given to BBONT by the Inland Revenue who received it in lieu of death duties.

Location
The reserve, though lying close to the edge of Oxford, is rather tricky to reach. Leave Oxford on the A40 for Witney. It passes below the A43 Oxford by-pass, and after about 1½ miles, you reach a layby on the left. Park here and walk back towards Oxford for about 1200 yards. Access is gained down the track. The River Thames forms the southern boundary of the reserve.

Access
Open to the public but please keep to the footpath.

Description
Oxey (once Ox Hay) Mead is part of a larger area known as Yarnton Meadows. When the A40 was built in the 1930's, the land to its north was sold and ploughed. Oxey was never ploughed and survived as the echo of a bygone age, continuing under traditional management of mowing for hay followed by grazing of the 'aftermath' until it came into BBONT's care.

The history of Yarnton Meadows dates back to Domesday when, as was usual, this village hay-land was held in common by the villagers. It was a lot meadow – each 'commoner' had rights to a strip of the meadow, and these strips were allocated in June when wooden balls carrying the commoners' names or marks were drawn from a bag. Once drawn, a swathe was cut from the strip and the new owners' initials cut into the ground with a knife. The luck of the draw gained the peasant his lot for that year, and his harvest of its slightly better (or poorer) hay.

We have here a close link between tradition and richly diverse nature. In a meadow such as this there can be as many as 75 different species of plant, including perhaps 15 different wild grasses. Amongst the galaxy of wildflowers are great burnet (flowering 6–9), pepper-saxifrage (6–8) and devils'-bit scabious (6–10).

There is also an abundance of grasshoppers, spiders and beetles. Curlew, snipe and redshank may perhaps be seen.

Management
Mowing at the end of June/beginning of July followed by cattle grazing until the end of November.
NOTE: Yarnton and Pixey Meads (alongside) are still owned jointly by local farmers with their rights in named cherrywood balls – Walter Jeoffrey, Harry, Gilbert, Boat, Watery Molly, William of Bladon – not their own names of course, but those of the owners of yesteryear. Today's owners have not met recently to draw lots but they may take up the old tradition. Agreements have been made with the Nature Conservancy Council to safeguard the wildlife interest of these two meads.

OXEY MEAD

OS sheet 164; SP 480107

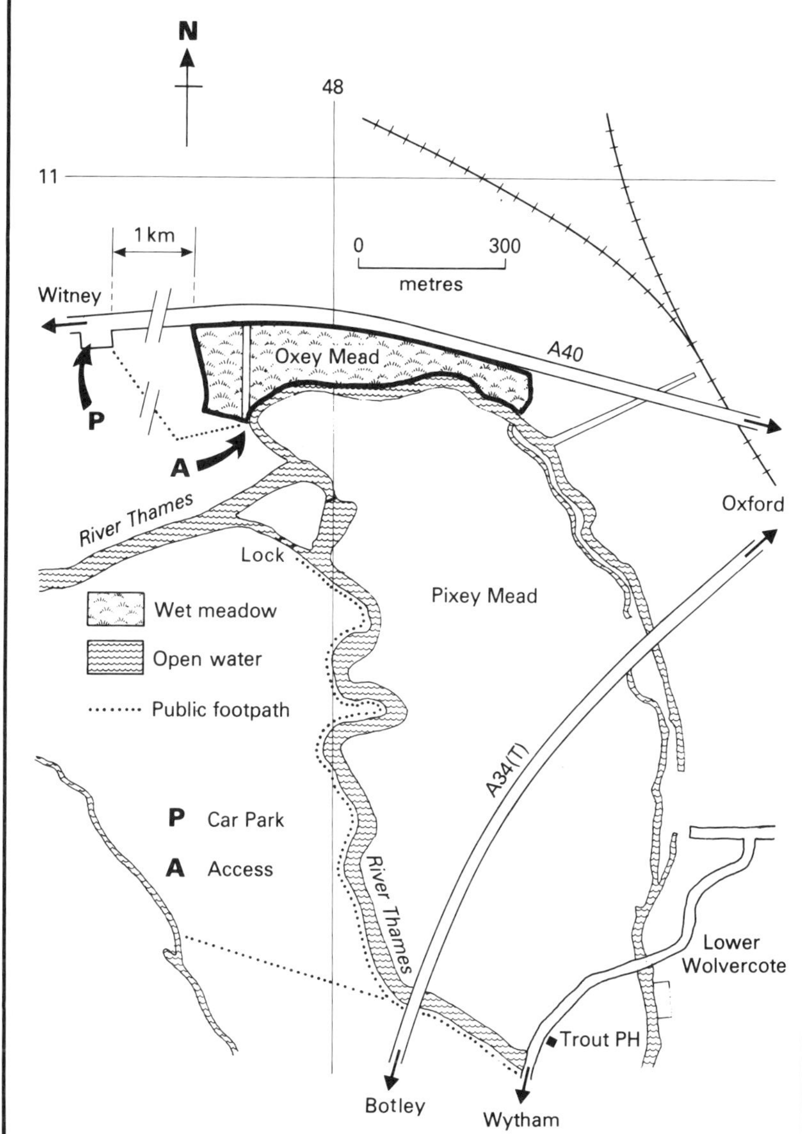

THE SLADE, BLOXHAM

OS sheet 151; SP 421357

Nearest towns Bloxham and Banbury

A mixed reserve including a stretch of disused railway, wet meadow and woodland. It covers 2.2 hectares (5½ acres) and is leased to BBONT by Oxfordshire County Council.

Location
From Banbury take the A361 towards Chipping Norton; go through Bloxham and on the southern edge of the town park on the verge by the playing fields on the right. The footpath leads from the gate across the playing field to a stile in the north-west corner. Continue through the next field along the northern boundary fence to the stile just past the last house; this leads into the reserve.

Access
Open to the public. The wet meadow and woodland are rather delicate habitats, so please keep to the pathway.

Description
The reserve has a good variety of habitats within a small area. Apart from the disused railway line, wet meadow and woodland, two small streams meet in the middle and flow out at the north-east corner of the reserve. Over 170 different plants have been recorded and the site is also good for birdlife and butterflies, although no very rare or restricted species are known.

One feature of interest is a small mound, the ruin of a ‘pest house’ serving Bloxham. A pest house was once a fairly common village feature – a cottage where those with contagious diseases could be isolated. Here it was surrounded by a small moat which healthy villagers dared not cross – requests for food and news were shouted across. Food was left on a special stone by the moat, to be taken in when the villagers had retreated.

This pest house was apparently in use into the early years of this century. In the smallpox epidemic of 1850, of all the ‘patients’ only one man came out alive.

Management
Work has included cutting parts of the wet meadow, maintenance of footpaths, fences and stiles and the creation of ponds. Woodland management includes coppicing and also pollarding willows.

The reserve is used by a local WATCH group who hold regular monthly meetings at the site to compile their nature diary. The children have collected a great deal of information and have a good understanding of how the reserve changes with the seasons. The WATCH group is also involved with the management work on the reserve, assisting with a great variety of tasks.

This is a good example of how reserves can be used to encourage children to learn about and help conserve wildlife habitats. Several BBONT nature reserves are used regularly by WATCH groups in this way.

THE SLADE, BLOXHAM

OS sheet 151; SP 421357

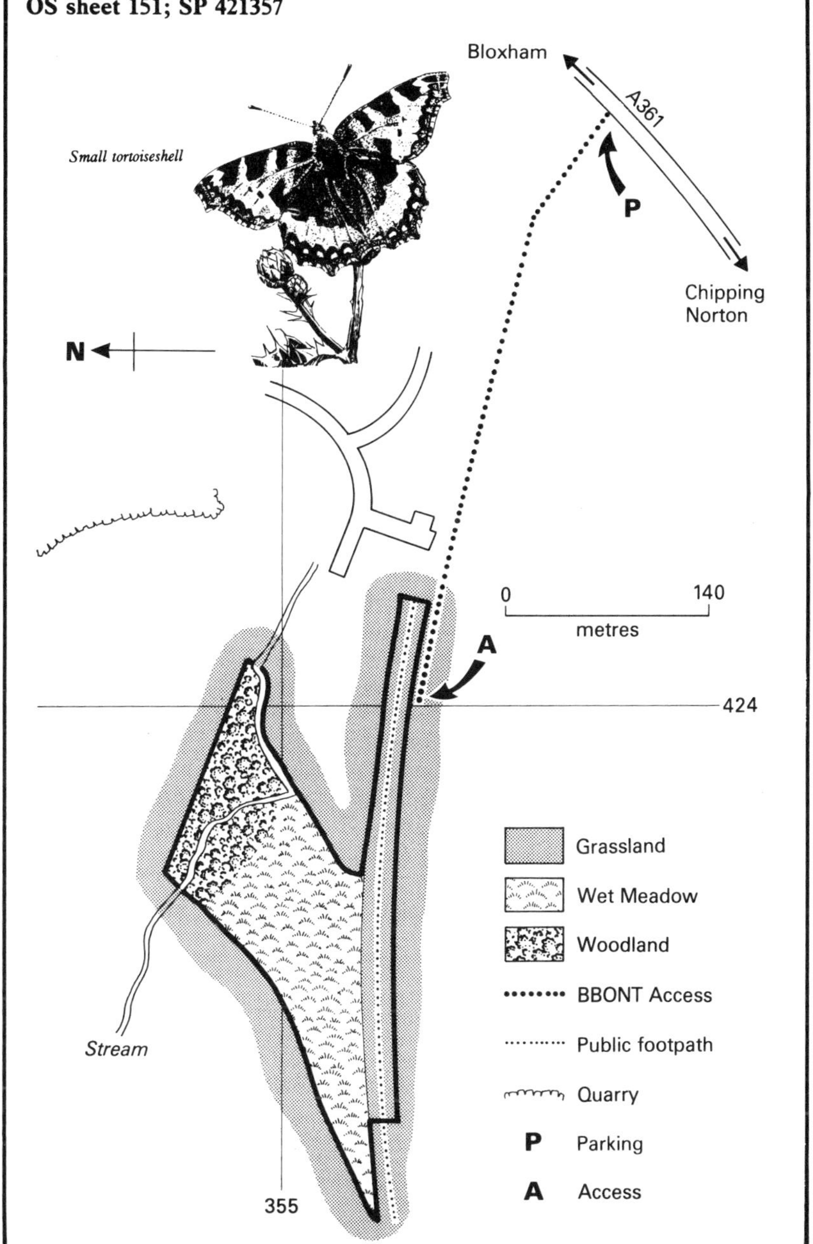

SYDLINGS COPSE and COLLEGE POND

OS sheet 164; SU 554095

Nearest town Oxford

A very varied reserve of 16.8 hectares (42 acres). Sydlings Copse was bought from Brasenose College in 1977 and College Pond is managed under an agreement with Christ Church, dated 1975. It is part of a larger SSSI.

Location
From the Headington roundabout on the ring road, proceed northwards past the crematorium and turn left on the B4027. Go about 350 yards to a point opposite Royal Oak farmhouse. Park on the road verge and walk southwards along a bridleway for about half a mile, past two pieces of woodland. Just past the second wood you will see another wood one field distant on your right. This is the reserve.

Access
Open to the public.

Description
The site consists of a mosaic of several habitats crammed in a steeply sided valley. They include reedbed, fen, carr, a stream, broadleaved woodland, coniferous woodland and heath. There are also more open areas of grassland which (because of the underlying geology) provide interesting comparisons. Some areas have an acid soil and here bracken and gorse can be seen, while the other grassy areas close by are over limestone and thus alkaline. This latter grassland is dotted with a myriad of wildflowers (over 120 have been recorded). Look out for large thyme (flowering 5–8), hairy violet (4), woolly thistle (7–9) and wild liquorice (7–8).

The broadleaved woodland is mainly oak and hazel – a traditional combination – while the joint presence of herb-Paris (5–7), wood spurge (4–5) and early-purple orchid (4–6) all point to it being old woodland.

The valley fen supports many uncommon wildflowers including bog pimpernel (6–8), marsh helleborine (7–8) and, unusual for a fen, columbine (5–7).

In all over 400 plant species have been recorded and it has for a long time been a favourite site for botanists from Oxford University.

As to be expected, the bird life is outstanding, while 28 different species of butterfly (including a good colony of marbled white) and 149 moth species have been recorded.

Management
The conifers were planted between 1941 and 1944 and have mostly grown well, only the larch planted into the limestone grassland have failed, apart from a few trees. Most of this larch has been removed by BBONT to favour the grassland. In addition most of the trees along the valley bottom of Sydlings Copse have been felled and the stumps treated. The Scots pines have been thinned and efforts are made to reduce the bracken in favour of other heathland species. Growths of willow and other wetland scrub encroaching into College Pond have been removed to increase the area of open fen (the original habitat). The limestone grassland is partly grazed annually. Coppicing of hazel on the valley sides has started again.

SYDLINGS COPSE and COLLEGE POND

OS sheet 164; SU 554095

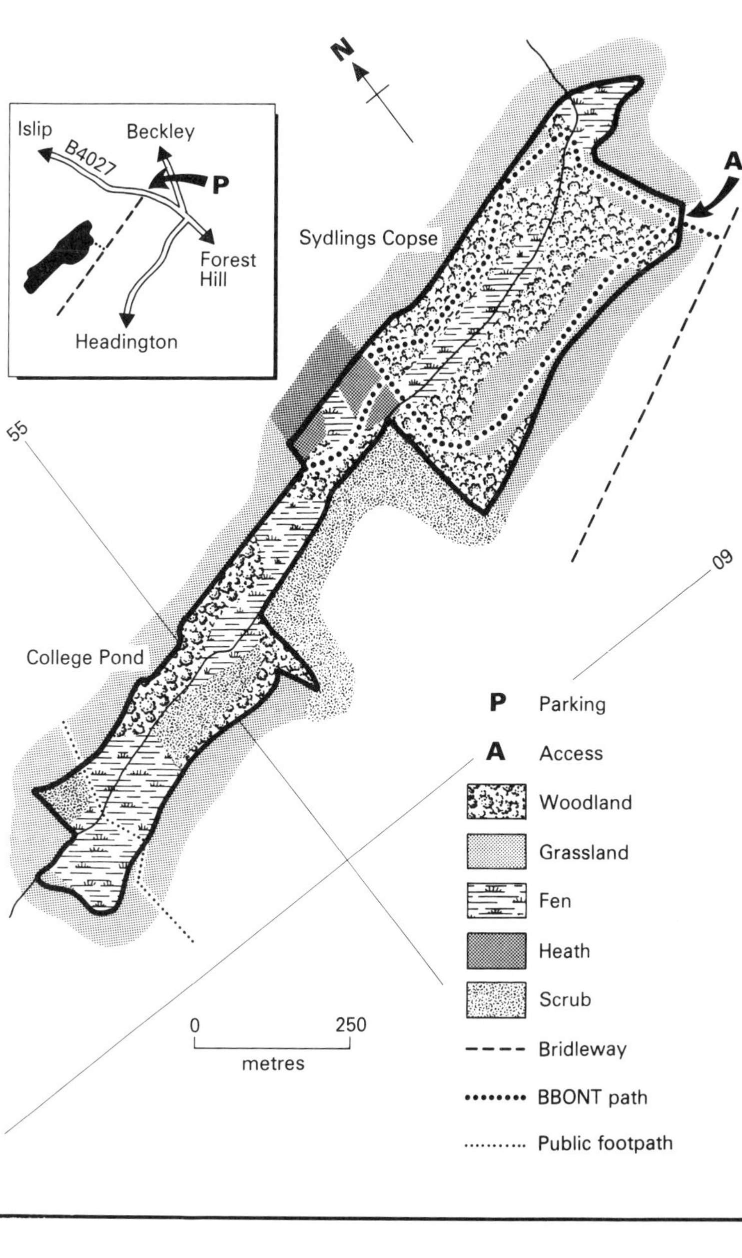

TUCKMILL MEADOW

OS sheet 174; SU 240900

Nearest towns Faringdon and Shrivenham

A meadow with a stream and a small spinney totalling 5.3 hectares (13½ acres). Because of the remnants of calcareous fen it contains, it is an SSSI. It was bought by the Vale of the White Horse District Council (who fenced the boundaries) and is on lease from them at £1 per annum.

Location
From Faringdon take the A420 towards Swindon; turn off left to Watchfield. Go through the village towards Shrivenham and as the road bends sharp left, turn right onto the track past Bremhill Golf Club. About 100 yards after passing the driveway to the Golf Club, there is a gate into a small car park, at SU 238897.

Access
Open to the public.

Description
The name 'Tuckmill' refers to a fulling mill which once stood on the northern bank of the brook. 'Tucking' is a local work for fulling, that is, beating cloth in a mixture of fuller's earth and water to increase its density. Fuller's earth is still dug locally at Baulking. Power from the water wheel lifted the drop hammers for the processing.

The reserve's main attraction is an L-shaped meadow plus most of the small spinney known as Ratcoombe Copse. The higher ground on the long arm of the meadow is limestone grassland with a good variety of wildflowers. The relic fen lies mainly alongside the stream in the long arm, marked by such typical plants as meadowsweet (flowering 6–9) and great willowherb (7–8). There are stands of reed in some places, bearing handsome tassels of flower heads in late Summer. The shorter arm is mainly marshy grassland (and one interesting feature of this part of the reserve is the old stream bed meandering across it). The wetter areas are good for dragonflies and damselflies (more needs to be known about the populations here) and kingfishers and herons have been seen. The brook forms part of the reserve.

Ratcoombe Copse lies on a slope above the fen and has a variety of shrubs, one being guelder-rose (6–7); it is, however, being invaded by sycamore.

The hedgerows around the meadow are also a rich feature, providing a good habitat for invertebrates and hence a larder for breeding birds such as chiffchaffs and spotted flycatchers. Moorhens, sedge and reed warblers and reed buntings are known to nest on the reserve.

Management
The grasslands have suffered in the past through lack of management. The site is now grazed by sheep and cattle and should recover its former interest in plants, butterflies and other insects. Ratcoombe Copse needs a great deal of attention and a programme is under way to remove all the invasive sycamore and some poplar and to replant with oak, ash, alder and other native trees. Use of the reserve for educational purposes is increasing and access is being improved.

TUCKMILL MEADOW

OS sheet 174; SU 240900

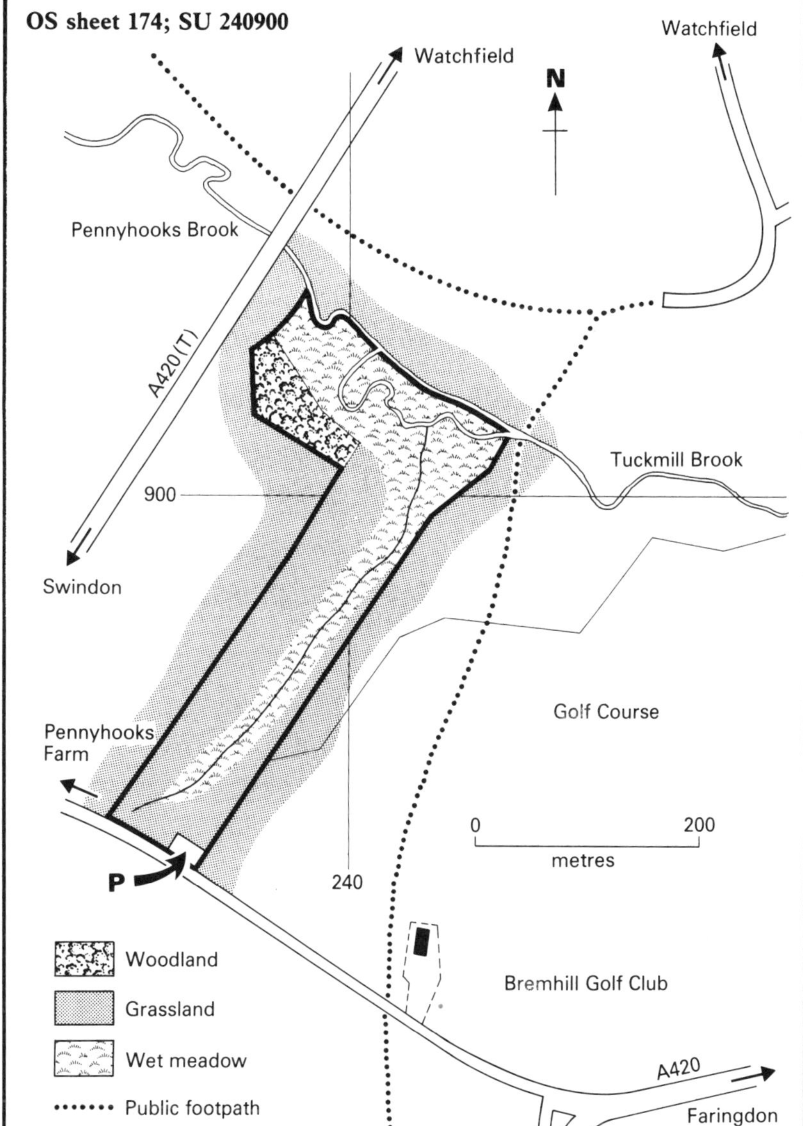

VICARAGE PIT

OS sheet 164; SP 401056

Nearest town Witney

A water-filled pit on lease from Oxfordshire County Council who declared it a Local Nature Reserve in 1976. The reserve totals 9.2 hectares (23 acres).

Location
The reserve lies south-east of Witney on the north side of the minor road which runs between Hardwick and Stanton Harcourt and is about 1 mile west of the last village. The car park is inside the gates at the south-east corner of the reserve.

Access
Open to the public, but please note that if interesting wildfowl are seen from the road, it is best to remain there if you want to watch them. If you enter the reserve these birds tend to leave, and fly to the neighbouring pits to the south. They are also disturbed when anglers are present (see under Management below).

Description
This water-filled pit has steep banks and deep water on its north, west and south sides. The eastern end is shallow and beds of reed are beginning to develop. The south-east corner of the reserve consists of a willow copse and here there is a vigorous colony of common spotted-orchids (flowering 6–8). The grassland areas show a good variety of wildflowers.

The reserve's main attraction, however, is the chance of seeing interesting waterfowl, including an impressive list of stray visitors. In the past these have included greylag geese and barnacle geese. The barnacle goose was, in monastic times, believed to hatch from goose barnacles which are found attached to timbers in the flotsam washed up along the coast. There was a distinct advantage in this curious belief for these geese, being so obviously of marine origin, could count as fish and be eaten on Fridays!

The snow goose has also been recorded here. It is a lone traveller, and when blown off course it joins flocks of other species, where its white plumage stands out amongst the greys and browns of its companions.

Management
Young willows have been removed from the gravelly areas in the north-east corner in order to create a suitable nesting ground for the little ringed plover, which did nest here in the past (it is attracted by wide expanses of bare ground, especially shingle, alongside open water). Some clearance of willows in the copse around the orchid colony has been undertaken. The grassy areas are mown in Autumn, after the Summer flowers have set seed. A hide is under construction on the south side of the pit.

The lake has been stocked with a variety of fish by the angling club which has private fishing rights.

VICARAGE PIT

OS sheet 164; SP 401056

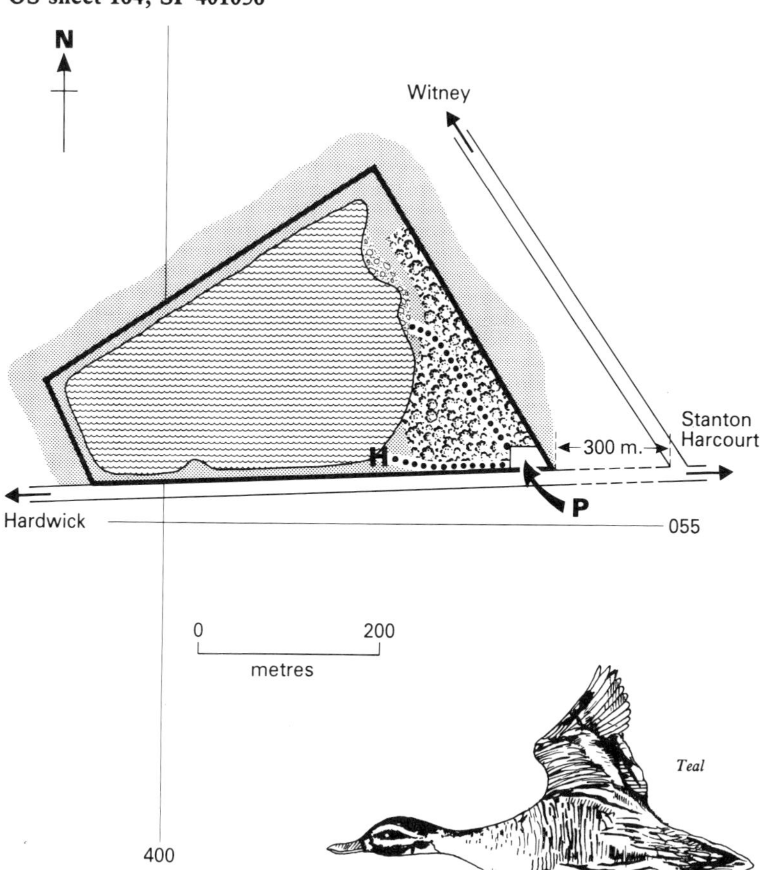

Open Water
Grassland
Woodland & Scrub
Shingle
H Hide
•••••• Footpaths
P Car Park

Teal

THE WARBURG RESERVE

OS sheet 175; SU 720880

Nearest town Henley-on-Thames

A complex of woodland and grassland, 103 hectares (257 acres) in area. It is an SSSI. 247 acres were bought by BBONT in 1967, the remainder later.

Location
Leave Henley-on-Thames north-westwards on the A423 and fork right at the end of the Fair Mile onto the B480. Turn left just north of Middle Assendon and follow the twisty lane for two miles through a quiet valley. Shortly after the lane becomes a track, you reach the reserve. The car park is on the right.

Access
Open to the public.

Description
This is one of BBONT's grandest and most prized reserves. It lies on both slopes of a dry, winding valley cutting through the chalk and encompasses a complex mixture of woodland and grassland habitats. There is a variety of soil types and the reserve has an interesting local climate, which may even produce unexpected frosts because of the pooling of cold air in the valley bottom. Over 450 different species of higher plants have been recorded, including 15 wild orchids and 50 plants which, in southern England, are usually found only in ancient woodland. Some of the woodland appears on maps of 1768, and these areas are likely to be ancient woodland, never cleared for fields. No less than 850 species of fungi have been listed, while 37 species of butterfly and 250 species of moth have been recorded. 34 different species of bird have nested here (and 75 species have been sighted in the reserve). Muntjac and fallow deer, dormice, adders, slow-worms, grass snakes and lizards are present.

Each of these species has its own habitat requirements, and many will be found in localised areas of the reserve only. In brief, the main habitats are mixed broadleaved woodland with the main trees being oak, ash, beech, birch, field maple and yew together with conifers from previous plantings; some beech woodland; conifer plantations of larch, spruce and Corsican pine; scrub; and open grassland.

The broadleaved woodlands have a rich variety of shrubs making up the understorey below the trees, while there are plentiful woodland wildflowers. On the grassy rides and open grassland areas grow a good many wildflowers typical of chalk soils.

Added to which, there is an imaginative interpretive centre and a nature trail for school parties. Trail guides and lists of the plants and animals are available.

Management
The aim is to maintain the variety of habitats while expanding the area of chalk grassland. Management includes coppicing the woodland, the removal of alien trees such as sycamore, Turkey oak and some of the conifers, and sheep grazing.

THE WARBURG RESERVE

OS sheet 175; SU 720880

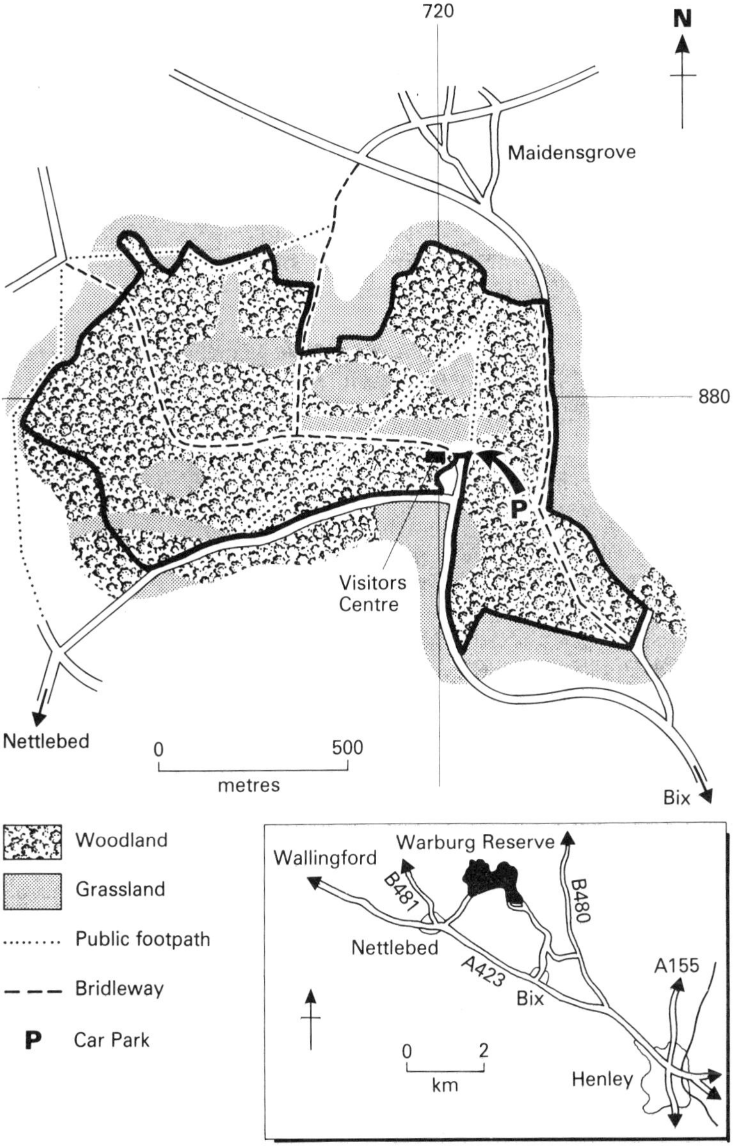

WARREN BANK

OS sheet 175; SU 653857

Nearest town Wallingford

An area of chalk grassland and scrub, 2.8 hectares (7 acres) in area. It is managed by BBONT in agreement with the owners dated 1972.

Location
Leave Wallingford to the east, cross the river and after ¼ mile turn right along the A4074 towards Reading. After 2 miles turn left at the White House Inn. Turn left at the crossroads and right after 100 yards through the hamlet of Hailey and on up a rough track. 100 yards after entering the wood, park on the right. Follow the track to the right to the field gate. Walk straight ahead for 200 yards keeping to the right of the wire dividing fence. A stile and notice board on the left mark the entrance to the reserve.

Access
Open to the public. The small plantation near the bottom of the slope is not part of the reserve, and the owners shoot there from time to time. Please delay your visit if a shoot is in progress.

Description
Here, surrounded by scrub, is a small relic of the chalk grasslands that once covered most of the downs. The grassland is divided into three paddocks and in it flower downland plants such as marjoram (flowering 7–9), cowslip (4–5) and bee orchid (6–7). Others typical of chalk downland to be seen here and in some other grassland reserves are field scabious (7–9) and chalk milkwort, with flowers (5–6) of a piercing gentian-blue.

The history of downland fragments such as this is interesting. When in flower, they can be as colourful as the stained glass in a village church, but they are in fact much older. When the Ice Age began to wane, trees invaded Britain eventually (around 8,000 years ago) creating a tangled wildwood in all but the wettest and bleakest places.

Eventually, the rising sea level from the melting ice caps flooded the English Channel, our first farmers, neolithic (new stone age) tribes, crossed into England bringing cattle, sheep, goats and pigs. They sensibly chose the lightly wooded soils such as gravel islands in the river valleys and the chalk downland for their settlements. They felled and fired the trees for clearings for skimpy harvests of corn and their animals grazed (and in time destroyed) the surrounding woodland. The downs became open.

Soil impoverishment and continual grazing meant that trees found it hard to recolonise the downland. Grazing and poor manuring (mainly by sheep) century after century encouraged small plants at the expense of large, which made up for their lack of size in the brilliance of their flowers.

Ploughing in recent decades has destroyed most of these grasslands.

Management
Some scrub has been removed and more needs to be taken out. The grassland is grazed by sheep.

WARREN BANK

OS sheet 175; SU 653857

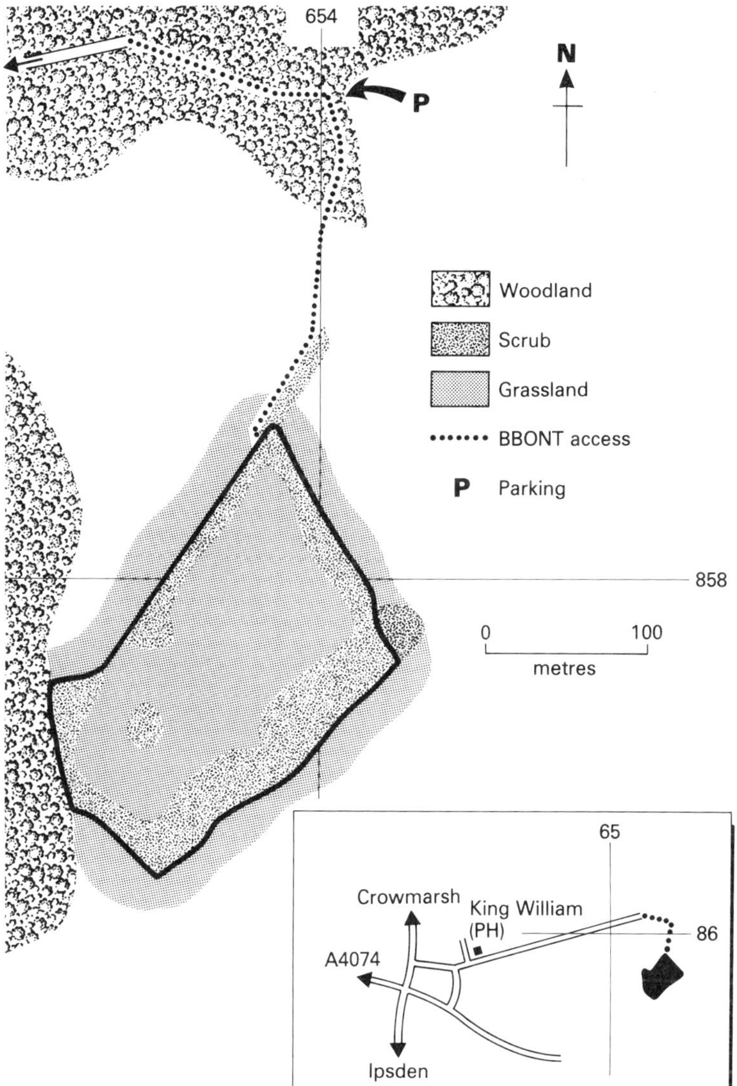

WESTWELL GORSE

OS sheet 163; SP 220114

Nearest towns Burford and Witney

A small area of grassland and scrub, 2.28 hectares (5.7 acres) in area. It is leased from the owners.

Location
Leave Burford westwards on the A40 for 1 mile, and turn left along the A433. After one mile park on the verge where a farm track leads off to the right. The reserve is on the right at this junction.

The road boundary of the reserve is marked by a Cotswold dry stone wall, which was restored by the British Trust for Conservation Volunteers (BTCV) in 1988.

Access
Open to the public.

Description
The reserve lies against the western boundary of the Oxfordshire–Gloucestershire border, and is set in the north-western corner of Westwell Parish.

The 1770 Enclosure Award set aside this area for raising furze and other fuel for the benefit of the poor of Westwell Parish.

It was originally divided into 7 sectors, one being cut each year for fuel. Although this practice came to an end in the middle of the 19th century, the land was then and still is administered by the Westwell Charities.

BBONT was interested in the site as an example of unimproved grassland among the surrounding intensively farmed arable fields. The underlying rock is limestone, the soil limey in character and a variety of wildflowers can be seen, including harebell (flowering 7–9), dropwort (5–8), pyramidal (6–8) and fragrant orchid (6–7) and common twayblade (6–7), field scabious (7–9) and the uncommon purple milk-vetch (5–7). In all, more than 200 different plant species have been recorded in this reserve.

There is a variety of butterflies and other insects.

The scrub area is also attractive to nesting and roosting birds.

Management
There was no maintenance prior to BBONT's lease, and as a result scrub had invaded the grassland. Current management includes pushing back the scrub to increase the area of grassland.

Rides cut through the scrub allow the visitor to wander through these parts of the reserve – the extensive lichen growths on the blackthorn are worth inspecting. These rides also provide sheltered open areas for butterflies.

Dutch elm disease in the 1970's destroyed the elm woodland which once covered about a third of the site on the western side. Regrowth of the elm is being encouraged (it can sucker healthily from its root system) and some new trees have been planted in the woodland area.

WESTWELL GORSE

OS sheet 163; SP 220114

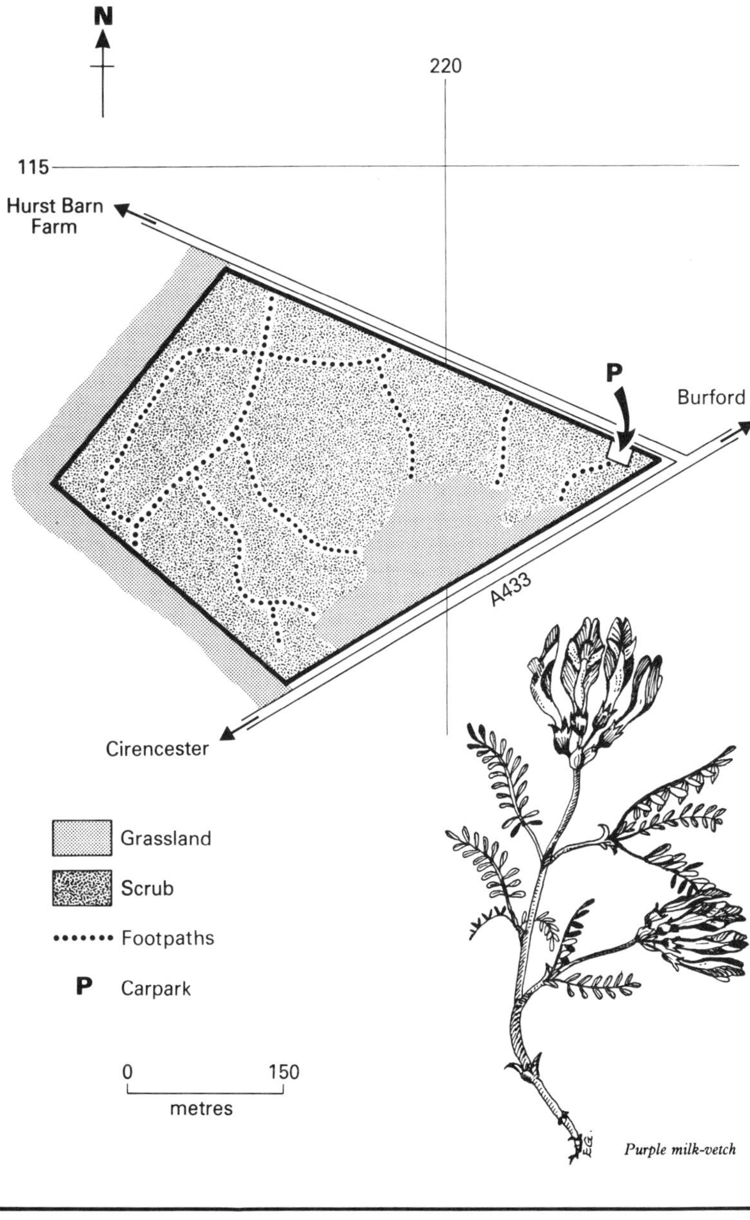

Purple milk-vetch

WHITECROSS GREEN WOOD

OS sheet 164; SP 603145

Nearest towns Oxford and Bicester

A 62 hectare (156 acres) woodland. It is an SSSI bought by BBONT in 1984.

Location
Leave Oxford northwards along the A43. Take the unclassified road to the right to Islip. Turn left in Islip towards Merton and take the first right through Oddington, Charlton-on-Otmoor and Murcott. One mile past Murcott there is a small cottage on your left, immediately opposite the reserve. Until the car park is completed, park on the hard-standing in front of the gate and **not on the grass** which belongs to the owner of the cottage. Please keep both gates shut.

Access
Open to the public along all rides. Dogs are not allowed.

Description
The reserve straddles the Oxfordshire/Buckinghamshire county boundary. The small part lying in Buckinghamshire formed part of the Royal Forest of Bernwood. The wood was shown on a map of 1590, and a large boundary bank and ditch can still be seen. It is an ancient woodland and, as to be expected, harbours a great number of plants – the total recorded at present stands at 200 different species.

Unfortunately, between 1963 and 1965, over half this old wood was felled and replanted with conifers by the Forestry Commission. However, the rides through the pines are still rich in wildflowers and busy with insects of all kinds. The remainder of the wood consists mainly of blackthorn/hawthorn thickets and overgrown hazel coppice with some oak and ash standards. Amongst the wildflowers to be seen here are bluebell (flowering 4–6), ramsons (4–6), yellow archangel (5–6) and enchanter's nightshade (6–8). After removal of the pines, the reserve will gradually be restored to native woodland. This will be a slow process, taking 50 years or more, but future generations will benefit from our efforts. Ecologists are already interested in studying the effects of BBONT's management on wildlife within the wood.

Mammals seen include fallow and muntjac deer. The many butterflies include black hairstreak, purple emperor, wood white and marbled white. The nightingale, turtle dove and grasshopper warbler nest here.

Management
The Scots pines will be thinned and eventually removed to provide a cash crop. The pine area will be returned to native high forest. Part of the neglected hazel is being coppiced to allow light to reach the woodland soil and encourage the growth of wildflowers associated with coppice woodlands. The rides are managed by mowing and grazing to control the growth of scrub and coarse grasses which would crowd out the less vigorous wildflowers.

BBONT has created further open areas by clearing 24 scallops along the central ride. Some will scrub up while others will be kept as grassland by grazing and mowing. The ponds are being cleared of silt and scrub.

WHITECROSS GREEN WOOD

OS sheet 164; SP 603145

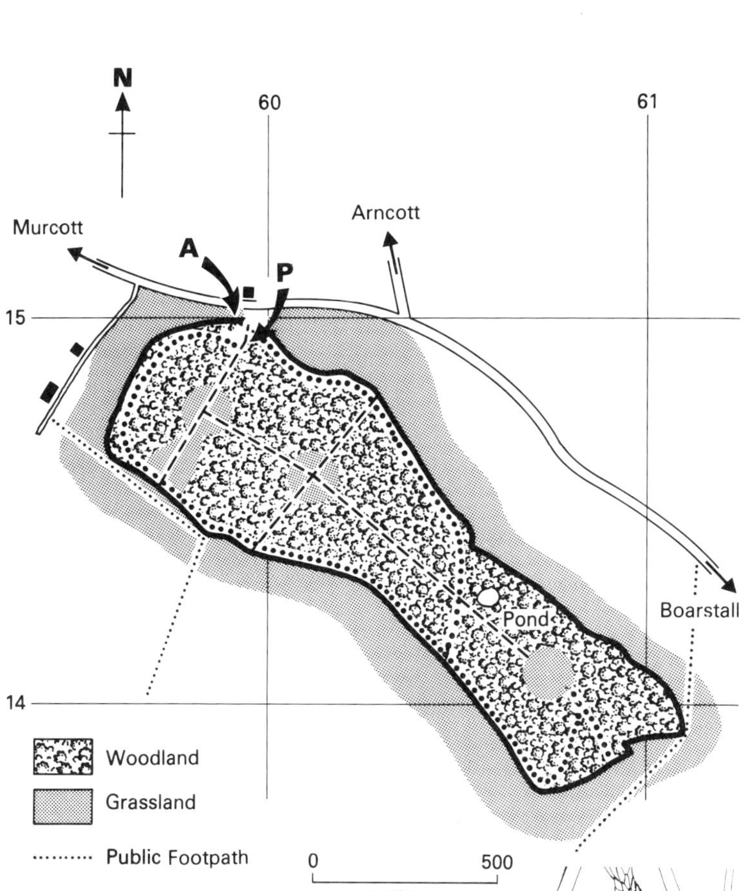

Wood white

WOODSIDES MEADOW, WENDLEBURY

OS sheet 164; SP 556178

Nearest towns Oxford and Bicester

An old hayfield of 3.5 hectares (9 acres). Part of a larger Grade I SSSI, it was purchased by BBONT in 1989.

Location
From the A43 between Oxford and Bicester, take the A4027 south to Islip. In Islip take the road for Merton. You pass roads for Oddington, and then after about 3 miles from Islip the road for Charlton-on-Otmoor joins on the right. Opposite is a farm road running past Home Farm. This farm road soon becomes very pitted, but must be endured for about ¾ mile. Park at the end, just before the railway.

The route to the reserve lies to the right, along the green lane. Follow this to the end where it turns sharply left. You will be faced by a pair of gates on either side of the railway, which should not be locked. Pay attention to the warning signs – though this is a byline, the railway is in use and trains must be expected from time to time.

Past the gate on the far side of the railway lies a track leading to the right, to the field which comprises the reserve.

Access
Open to the public.

Description
One of a series of exceptionally rich wildflower meadows, the wildlife value of which was not discovered until the mid 1970's. A detailed survey of this particular meadow has not yet been undertaken, but wildflower meadows such as these are a speciality of the Oxfordshire countryside. A riot of colour in Spring and Summer, these meadows hum and buzz with a wide variety of insect life.

Notable plants to look for include cowslip (flowering 4–5), cuckooflower (4–6) and green-winged orchid (5–6).

Green-winged orchid

Management
It is likely to be cut for hay and the aftermath grazed.

WOODSIDES MEADOW, WENDLEBURY

OS sheet 164; SP 556178

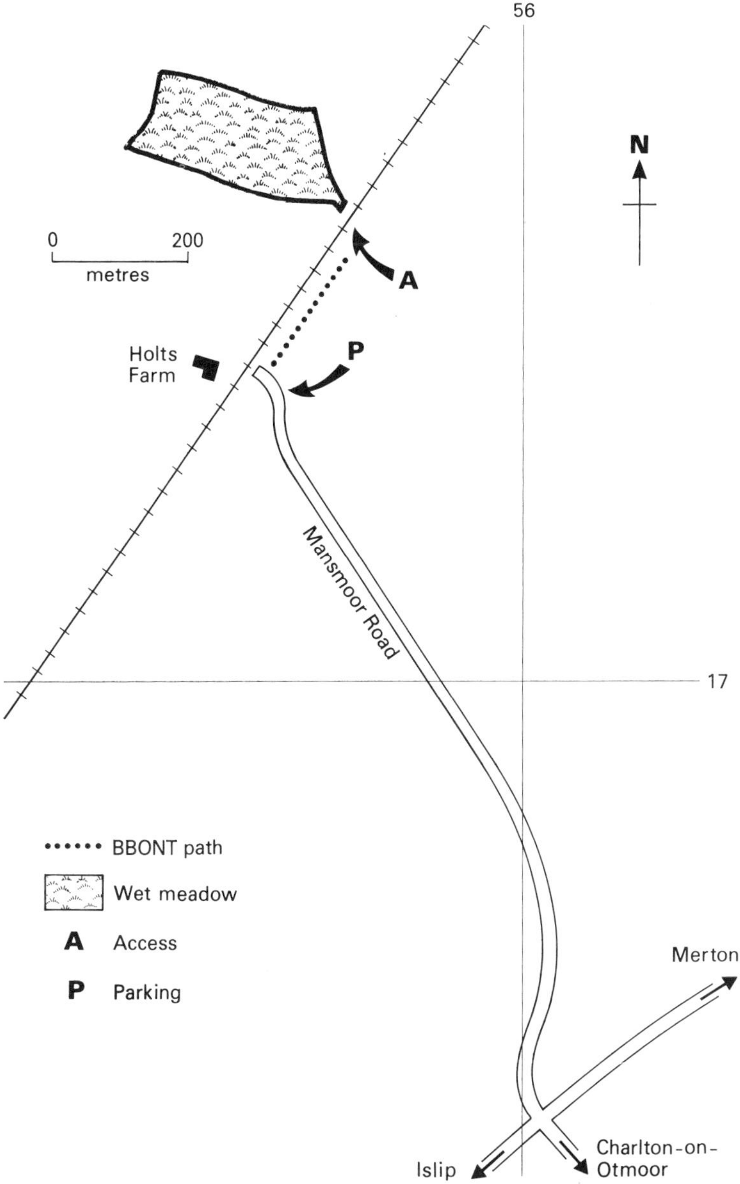

PERMIT ONLY Reserves

Ardley Quarry
Aston Upthorpe
Calvert Jubilee and Extension
Crog Hill
Dancersend Waterworks
Eversdown
Foxcote Reservoir
Hollington Wood
Hollybank Marsh
Iffley Island
Lodge Wood
Munday Dean
Otmoor Rifle Range and Spinney
Parsonage Moor
Pitstone Fen
Rack Marsh
Robertswood
The Spinney, Shotover
Swains Wood

ARDLEY QUARRY

Nearest towns Bicester and Banbury

A quarry, part of an SSSI, and managed by BBONT by agreement with the owners, dated 1983. The site occupies 12 hectares (30 acres).

Access
By permit only, obtainable from the BBONT Office in Oxford. The quarry lies directly under the approach to Upper Heyford Airfield and the noise of the planes can be startling (it does not have much effect on the wildlife, however).

Description
This is a very interesting site in many ways. The quarry and the railway cutting with which it is associated are of great geological interest. Details are for the dedicated (and can be obtained with the permit) but it is a key site for the interpretation and correlation of the layering of rocks of the Jurassic era.

A quarry with its bare rock faces is always a satisfying place to visit. Apart from its geology, the site has some historical interest too as there are medieval earthworks to be seen. From a wildlife point of view, there are different habitats – old spoil tips, scrub, woodland and a pool below the rock face which dries up on occasions. There is much rough grassland carrying a good many lime-loving wildflowers, such as cowslips (flowering 4–5), various vetches, and wild orchids which are at their best in June.

Sheltered areas such as this are usually good for butterflies and 30 species have been recorded. Along the cutting, for example, there is a large colony of the small blue butterfly (which lays its eggs on the abundant kidney vetch). There are also thriving colonies of grizzled skipper, dingy skipper and green hairstreak butterflies on the reserve.

35 species of birds nest here and there are lizards, snakes and many small mammals amongst the wildlife populations.

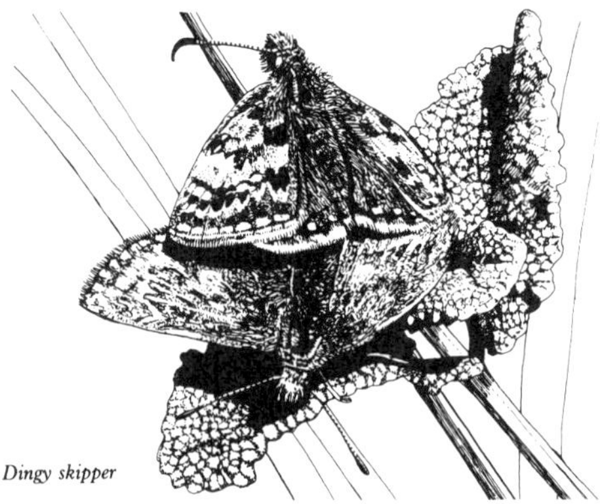

Dingy skipper

Management
A considerable amount of scrub clearance has been put in hand but more is waiting to be done.

ASTON UPTHORPE

Nearest towns Wallingford and Didcot

This grassland reserve of 16 hectares (39½ acres) is managed by agreement with the owner. It is part of a dry valley in the chalk downs which is registered as an SSSI.

Access
Strictly permit only, obtainable from the BBONT Office in Oxford.

Description
The reserve is an important remnant of flower-rich old downland, most of which has been lost to intensive arable farming. The pasqueflower is seen (flowering 4–5) and other plants of interest include dropwort (5–8), horseshoe vetch (5–7), small scabious (7–9), field fleawort (5–6), chalk milkwort (5–6) and burnt orchid (5–6). There is also an extensive area of juniper scrub.

The turf has to be grazed regularly to allow these plants to flourish. It is only a few years ago that enthusiasts hoping to protect the pasqueflower put a fence around the site to prevent disturbance. The fence also prevented the livestock grazing; the grasses grew tall and overwhelmed the plant.

As to be expected, there are many butterflies, including the chalkhill blue (which relies on the presence of the horseshoe vetch) and small blue (kidney vetch).

Amongst the birds, the short-eared owl has been recorded in the Winter months.

Pasqueflower

Management
Cattle belonging to the owner are grazed; they are removed during the Spring and early Summer.

CALVERT JUBILEE and EXTENSION

Nearest towns Bicester and Buckingham

A disused claypit, 34 hectares (87 acres) in area, plus a length of disused railway 3 hectares (7.4 acres) in area. Calvert Jubilee is on lease from the London Brick Company until 1992. The length of railway line along the northeast boundary was bought by BBONT in 1981.

Access
By permit only, obtainable from the BBONT Office in Oxford.

Description
This is a disused claypit whose northern end was used as a municipal tip and was covered with poor quality clay known as callow. But even here the natural world can produce surprises – notably in the form of wild orchids. Visitors can see the common spotted-orchid (flowering 6–8) and bee orchid (6–7) as well as blue fleabane (7–8), looking rather like a blue, but large, groundsel.

The lake area, of 50 acres, is a Winter sanctuary for wildfowl, with bird counts running well into four figures. Species seen include mallard, our most familiar duck and one that all can recognise, tufted duck and pochard. Tufted duck are often to be seen mingled in with the 'rafts' of pochard. The site is renowned for its gull roost. Wildfowl apart, many birds nest in the scrub and wooded areas, including nightingale and tree creeper. Green and great spotted woodpeckers are seen.

Thanks to the generosity of the London Brick Company, there is a bird-watchers' hide designed primarily for disabled visitors.

Tufted duck

Management
Thanks to the goodwill and generosity of the London Brick Company, a number of major works have been undertaken including the creation of three islands to enable waterfowl to nest safe from the local foxes (which are abundant). In addition, a small lagoon has been created at the north end of the lake in the hope of attracting frogs and other amphibians. The shallower areas at the northern end have already been planted with reed, to create dense reed beds.

CROG HILL

Nearest towns Wantage and Lambourn

A small remnant of chalk downland, 1.6 hectares (4 acres) in area. BBONT manages it by agreement with the owner dated 1980.

Access
By permit only, obtainable from the BBONT Office in Oxford.

Description
A small but interesting relic of old downland, this reserve consists of a long narrow bank between a farm track and race training gallops. Although small, a marvellous variety of plant species are seen – the tally is 141 to date. Wildflowers include autumn gentian (flowering 8–9), many different vetches and six different species of wild orchids (best in June). Butterflies include the chalkhill blue and Duke of Burgundy. The marsh fritillary has also been recorded. Despite its name, this last is not confined to marshy ground; its caterpillar food plant is devils'-bit scabious which grows in a variety of places.

Chalkhill blue

Management
In the past management has largely consisted of mowing the bank each year, but since 1986 the site has been sheep grazed which has proved even more beneficial.

DANCERSEND WATERWORKS

Nearest towns Wendover and Tring

A 4.3 hectare (10½ acre) site which is still an operating waterworks supplying Tring, Aston Clinton and Halton. It includes chalk pits and a lagoon with an earth dam and round concrete reservoirs. BBONT advises on the management of the site and carries out management work in a 5 acre area of old chalk pits.

Access
By permit only, obtainable from the BBONT Office in Oxford.

Description
These waterworks were constructed in 1886 to supply water to the various houses of the Rothschild family in the area. In 1940 they were taken over by what is now Thames Water. Most of the area was disturbed many years ago by the excavation of a number of chalk pits and a lagoon with an earth dam. Subsequently a series of circular concrete reservoirs were constructed with earth banks built up round them. These reservoirs are now disused, having been replaced by a larger covered reservoir. There is a large area of bare chalk soil which is still being colonised by plants.

Over many years, distinctive plant and animal communities have developed. More than 160 plant species have so far been identified, including orchids such as the white helleborine (flowering 6–7), common twayblade (6–7), pyramidal (6–8), fragrant (6–7), greater butterfly- (6–7), bee (6–7) and fly (5–6) orchids.

Management
The plant and animal communities have developed naturally over the decades. The lagoon and pits need to have the invasive scrub and trees removed for the benefit of the ground plants. A careful mowing plan is being developed for the large grassland area around the reservoirs; it will ensure that coarse growing species are controlled whilst finer flowers are allowed to bloom and set seed. Taller vegetation is to be left in places, to the benefit of grasshoppers, and also to benefit the brown, skipper and other butterflies and invertebrates which are linked with long grass.

EVERSDOWN

Nearest town Henley-on-Thames

A sizeable area (5.6 hectares, 14 acres) of chalk grassland bank, managed since 1980 by agreement with the owner.

Access
By permit only, obtainable from the BBONT Office in Oxford.

Description
The bank has become much overgrown with scrub, some of which has already been removed. It was known to have been fairly open grassland (at least in parts) in the 1960's but was rapidly invaded by scrub in the 1970's. Perhaps the decimation of rabbits by myxomatosis in the 1950's was in part the reason. Rabbits will keep grassland free of even hawthorn and blackthorn. Elderberry they find distasteful, however, and so this shrub often marks their warrens.

There are well-developed anthills, showing that the site has not been disturbed by farm machinery for many decades and perhaps never. These anthills are created by the meadow ant and are enlarged year by year, so that their size can be an indication of their age.

Over 150 different plant species can be seen, and butterflies in abundance. The scrub attracts a good many nesting birds and the berries and cover invite thrushes and other birds to feed and roost in Winter.

Common rock-rose

Management
Some of the more open scrub has been removed and the sward grazed by sheep. We hope to extend this open area. A swathe of scrub was removed when a gas pipeline was laid across the reserve and this area has been re-seeded with sheep's fescue grass, a wild grass typical of this type of soil. Wildflowers from the undisturbed grassland alongside are now seeding themselves into this area.

FOXCOTE RESERVOIR

Nearest town Buckingham

A reserve of 22 hectares (56 acres) comprising a reservoir and varied surrounds. It is an SSSI, held on licence from Anglian Water for 21 years from 1971.

Access
A hide to the south of the reservoir is open from time to time during the Winter – and visits are restricted to those days. Please ask for details from the BBONT Office in Oxford.

Description
The reserve includes the reservoir, which was completed in 1956, and the lower parts of the fields alongside the water. The aim is to provide an undisturbed wildfowl refuge – hence the restriction on visits. The sight of human beings in the surrounding fields or on the dam starts a panic amongst the birds and while some may retreat to the opposite bank, many are likely to fly off altogether.

Up to a thousand wild duck of various species may be seen here. It is a good site for seeing passage migrants. Wildfowl and waders flying north in Spring are usually in some haste to gain the better breeding sites on the edge of the Arctic. The return journey in late Summer and Autumn is likely to be more leisurely, however, and in the case of many species the adults are seen in advance of the juveniles, maybe by as much as a couple of weeks or more.

Mallard

Management
The hide is maintained, and it is important to maintain the screen leading from the gate to the hide. The vegetation on the bank to the west of the hide needs occasional mowing.

HOLLINGTON WOOD

Nearest town Milton Keynes

A bluebell wood of nearly 20 hectares (50 acres). It is managed by agreement with the owner who lives on site in what used to be the woodman's house. Although within BBONT's boundary, it is in fact managed by the neighbouring Trust – the Beds and Hunts Wildlife Trust.

Access
By permit only, obtainable from the BBONT Office in Oxford.

Description
This is a typical ancient Midland clay woodland, dominated by oak, with hazel featuring strongly in the understorey.

It was coppiced during World War I and many of the regrown trees are now gigantic.

There is a delightful stream running through the wood.

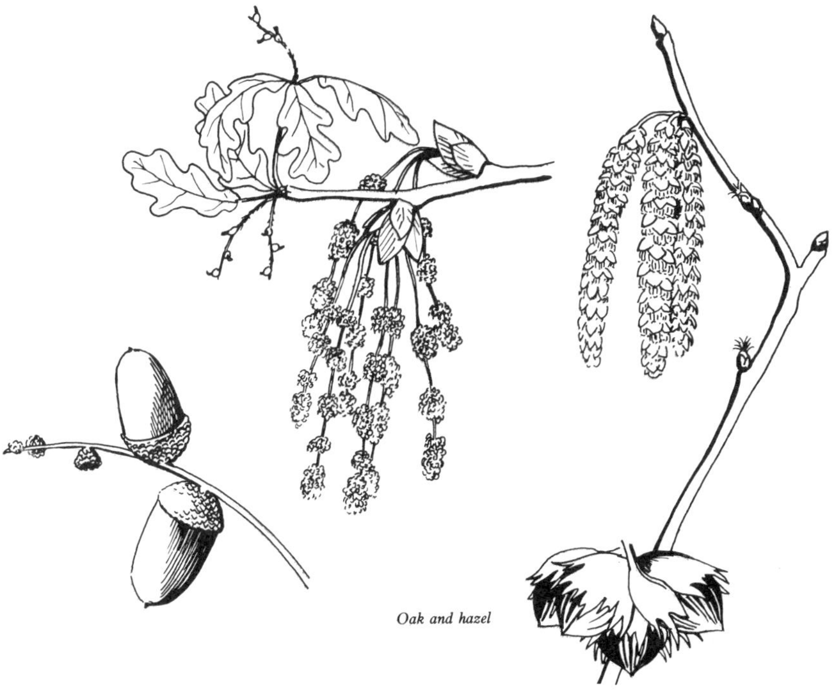

Oak and hazel

Management
This incorporates woodland 'edge' creation along the rides, coppice cutting, high forest management (with occasional thinning to open up the canopy) and areas that are left to themselves. Raising the wetness and creating a pond by damming the stream has been a recent task.

HOLLYBANK MARSH

Nearest town Oxford

A marshy meadow, 1.6 hectares (4 acres) in area, which BBONT has managed by agreement with the owner since 1978.

Access
By permit only, obtainable from the BBONT Office in Oxford.

Description
This marshy meadow is bordered on two sides by a small river. There is a small but spreading reedbed. There are colonies of southern marsh-orchid (flowering 6–8), and yellow iris is present (flowering 5–7).

Reed warblers have been recorded as breeding successfully and sedge warblers recorded as Summer residents, probably also breeding.

Sedge warbler

Management
Willows are pollarded and reeds will be cut occasionally to allow light to reach the lower growing plants. The spread of the reedbed is being monitored.

IFFLEY ISLAND

Nearest town Oxford

A peninsula about 0.1 hectare (¼ acre) in area. It is owned by the Oxford Preservation Trust and leased to a local resident who manages the site with advice from BBONT.

Access
This is by boat, and visits must be arranged well in advance through the BBONT Office in Oxford. Full details will be given at that time.

Description
The reserve is not actually an island, but a small peninsula between the main stream of the Thames and a backwater with the third side fenced. Fritillaries grow here (flowering 4–5), but they were being picked and by 1972 numbers were down to a mere 8 flowers. Now, thanks to the protection given by the reserve manager and a few helpers, numbers have increased to over 1,000.

Apart from these flowers, the reserve also provides a haven for many mallard, moorhens and coots. Mute swans regularly nest here, and kingfishers and herons are seen.

Fritillary

Management
Apart from fence maintenance, management includes the control of bindweed and the mowing and removal of the luxuriant growth of grass.

LODGE WOOD

Nearest towns Reading and Wokingham

A small wet woodland and strip of meadow, 1.8 hectares (4.5 acres) in all, part of the Lodge Wood and Sandford Mill SSSI. It has been held on lease from the owners since 1982.

Access
By permit only, obtainable from the BBONT Office in Oxford.

Description
The reserve consists of a small (around 2½ acre) wet woodland and its boundary ditches together with a strip of meadowland which runs along two of its sides and extends to the River Loddon. The soil is poorly drained and the wetter ground within the wood is marked by alder and willow but it is drier to the south with oak and ash and a few exotics planted. It is an old wood, being shown on a map of Berkshire of 1761.

Red currant (a native shrub) can be found in the wetter soils and bluebells (flowering 4–6) on the drier soils beneath the oaks. Here wild daffodil (4) also flowers; there are other wildflowers and many different mosses.

The strip of meadowland is important because it safeguards ditches on which the Loddon lily relies. This is now a very rare plant, in England confined to the Thames basin, although it is also seen as a garden flower. It gains its name from the River Loddon, and this site near that river is now one of its main strongholds. It has another name, however, the 'summer snowflake' – less logical, because it flowers in Spring, in April–May!

It is a wet-loving plant. Its fruits develop float chambers and in the past the frequent uncontrolled floods carried them for miles to lodge and grow in streamside woods. Floods would also carry the bulbs to new sites.

Summer snowflake or Loddon lily

Management
The aim is to maintain the damp conditions needed by the Loddon lily. The fields to the west, north and east have planning permission for extraction of sand and gravel. A lake will result and the Summerleaze Gravel Company will lease this to BBONT as a nature reserve. Apart from the waterfowl and other interest, once this lake is filled with water the future of the now very rare Loddon lily should be assured.

MUNDAY DEAN

Nearest town Marlow

A small paddock of 1 hectare (2.5 acres), bought by BBONT in 1980.

Access
By permit only from the BBONT Office in Oxford.

Description
The prime interest of the reserve is a colony of green-winged orchid (flowering 5–6), a flower which was a common sight in many meadows before the introduction of intensive farming in recent decades. The rich sward includes other early-flowering hay meadow flowers including the oxeye daisy (5–9) and yellow rattle (5–6). This last gains its name from the sound of the loose ripe seeds rattling around in the dry seed pods, and in olden times this was taken to be a warning that the hay was ready to cut.

There are a number of different woody shrubs in the lower hedge, indicating that it could be of some age, and towards the end of Summer a good number of different plant galls are to be found along this hedgerow. Bittersweet and black bryony are also be be seen in the hedges.

Green-winged orchid

Management
The field is managed as a traditional, unimproved pasture. The grasskeep is let, and the lessee mows the meadow each year after the orchids have flowered and set seed. Later in the year, the aftermath is grazed briefly by his livestock.

OTMOOR RIFLE RANGE and SPINNEY

Nearest town Oxford

Wet meadowland, some woodland and scrub totalling 28 hectares (70 acres). BBONT has managed the Rifle Range on licence since 1972 and the spinney has been managed by agreement with the owner since 1974. The reserve lies wholly within the Otmoor SSSI.

Access
Because of the active use of the Rifle Range, access is by permit only, obtainable from the BBONT Office in Oxford.

Description
Otmoor is a large bowl-shaped area on the flood plain of the River Ray and the SSSI is the core of what was until recent times an extensive area of wetland which was flooded in Winter and traditionally managed as rough grazing marsh. Wildfowl and waders abounded. In recent times, however, much of the land outside the SSSI has been drained for arable or improved pasture.

This reserve retains an exciting variety of plants and insects. Wildflowers include dyer's greenweed (flowering 6–8), ragged-robin (5–7), saw-wort (7–9), meadow thistle (6–7), sneezewort (7–8), devils'-bit scabious (6–10) and marsh stitchwort (5–8).

The insects number a strong colony of the endangered marsh fritillary butterfly (which lays its eggs on the devils'-bit scabious), the black hairstreak butterfly and the spectacular emperor moth.

The spinney is dominated by oak, planted onto once open land (the pattern of ancient ridge and furrow can still be seen beneath the trees). In spite of the disturbance from the rifle range this has a flourishing rookery and several pairs of nightingales nested and sang here in 1988.

Marsh fritillary

Management
Various parts of the rifle range are either grazed or mown for hay. The spinney, apart from the creation of a path, has had little management although there has been some path clearance along its western edge.

PARSONAGE MOOR

Nearest towns Abingdon and Oxford

The reserve of 12 hectares (30 acres) is part of the Cothill Fen SSSI. BBONT manages it by agreement with two separate owners dated 1980 and 1981.

Access
By permit only, obtainable from the BBONT Office in Oxford.

Description
This is part of the 55-acre Cothill Fen SSSI, one of the best fenland sites in the county. It has outstanding and now nationally rare calcareous (limey) fen and other wetland habitats. Altogether the SSSI shows succession from open water to fen and carr (wet woodland) with an area of ancient woodland adjacent.

300 different plant species have been recorded on the SSSI – a very large number in relation to its size. There is added interest in the fact that both lime-liking (calcicole) and lime-hating (calcifuge) species are seen. This is the result of the complex profile of the area, which sees grits covered by peat and marl and fed by rather limey water.

BBONT's reserve encompasses most of the open fenland and extensive reedbeds and also includes some scrub and woodland. There are many wildflowers to interest the visitor, such as grass-of-Parnassus (flowering 7–10), common butterwort (5–7), marsh helleborine (7–8), and dyer's greenweed (6–10).

Being so close to Oxford, this interesting place has a long history of research. A long-term study on the population dynamics and genetics of the scarlet tiger moth has become famous. The reserve is very rich in moths and dragonflies. It is also a fine place for birds, especially migrants in Winter.

Grass-of-Parnassus

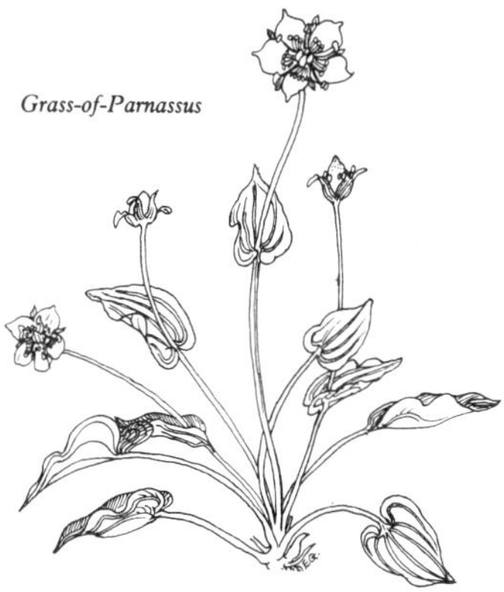

Management
Cothill Fen is a National Nature Reserve managed by the Nature Conservancy Council. Parsonage Moor itself is suffering from the invasion of scrub of various kinds and this is being removed.

It is hoped that a part of the fen area can be excavated with the intention of studying the succession from open water back to fen.

PITSTONE FEN

Nearest town Tring

A 3.6 hectare (9 acre) reserve created in 1981 by the Tunnel Cement Company (now the Castle Cement Company) and managed jointly with BBONT.

Access

All visitors must have a permit. The site is open to the general public on Sundays, 10am to 5pm, with permits from the Warden's Office, College Lake Wildlife Centre (see page 74) . BBONT members can visit at any time with a monthly or annual permit from the same office or from the Conservation Officer, Castle Cement (Pitstone) Ltd, Pitstone, Leighton Buzzard, Bedfordshire, LU7 9AR (please include a stamped addressed envelope).

Description

This is a half mile strip of land along a railway line which encompasses four distinct habitats. The first is a small Scots pine plantation, the second a strip of rough grassland, the third a fen running parallel to the grassland but some 20 feet below it, where chalk was quarried some 40 years ago. Finally, until 1963, flue-dust from the cement works was dumped in a ridge down the length of the fen and provides an unusual potash-rich habitat.

A total of about 250 different plant species have been recorded so far and 15 species of dragonfly and damselfly, together with 30 of butterflies, 300 of moths. Reed warblers breeding in the small reedbed are amongst the 100 bird species recorded. 14 different mammal species have been recorded and 4 amphibians.

All in all a very interesting reserve.

Common spotted-orchid

Management

Management includes control of scrub, grassland and reedbeds.

RACK MARSH

Nearest town Newbury

An area of meadow and wetland, 13.75 hectares (34 acres) in extent. It is managed on licence from the owner, signed in 1983.

Access
By permit only, obtainable from the BBONT Office in Oxford.

Description
Wetland such as this was characteristic of the Lambourn river valley before modern drainage and ploughing. A thick layer of peat has developed on top of the deposits of alluvium and gravel which the river has spread over the chalk. In this peat the remains of a prehistoric canoe have been discovered, the wood preserved by the waterlogged conditions.

The site has a history of use: there was a mill at the west of the reserve registered in the Domesday Book, 1086, and the name 'rack marsh' comes from the racks used to dry cloth when the site later became a fulling mill. The mill remains today as the Watermill Theatre.

The ground itself, however, has been little disturbed. There is a range of plant communities on the peat – marshy and other grassland, swamp, scrubby alder 'carr' and scattered willow scrub. The greatest diversity of plant life is seen in the marshy grassland. Early marsh-orchid (flowering 5–7) and southern marsh-orchid (6–8) are to be seen here, together with marsh valerian (5–6) and bogbean (5–7) amongst them. This last will interest the visitor, for it is counted one of our most handsome wildflowers.

As for birds, reed and sedge warblers are attracted to the reserve, reed bunting and moorhen may also nest, while heron and kingfisher are often seen.

Bogbean

Management
The site seems to have been grazed or cut for hay since early times. The western area has been fenced and will be grazed; a grazing or cutting regime is necessary to prevent the vegetation drying out and becoming rank.

ROBERTSWOOD

Nearest towns Tring and Aylesbury

A beechwood on a south facing slope of 15 hectares (37 acres), held on lease from the owner since 1976.

Access
By permit only, obtainable from the BBONT Office in Oxford.

Description
This south-facing chalk slope overlooking Wendover Dean was originally grassland, but was planted with beech together with a nurse crop of conifers at the end of World War II. The conifers have now been removed.

Beech woodland is normally rather poor in plant species, as the gloom created by the foliage and the long-lying leaf litter provide rather hostile conditions. Here, however, there is a remarkable display of yellow bird's-nest (flowering 7–8), a plant that can tolerate these conditions. It lives on rotten beech leaves and needs no green colouring to survive; the drooping, bell-shaped flowers, stem and scale-like leaves are all pale yellow. Over 200 species of other flowering plants have been recorded.

Yellow bird's-nest

Management
To maintain the variety of wildflowers, it will be necessary to open and maintain some large glades amongst the ranks of the growing beech trees. Some small glades have already been created. In addition the beech is being thinned and ash and sycamore which have seeded themselves in are being removed. A small pond has been created with an observation hide alongside.

THE SPINNEY, Shotover

Nearest town Oxford

An area of mixed woodland, 11.2 hectares (28 acres) in extent. It has been managed by agreement with the owners since 1982.

Access
By permit only, obtainable from the BBONT Office in Oxford.

Description
This is a mixed deciduous wood with areas of conifer plantation. A small stream runs through the wood along which, in the past, a number of artificial ponds have been created for amenity purposes. These ponds used to be stocked with trout but none now remain.

A path follows the valley alongside the ponds and there are open rides through the stands of conifers. As well as primroses (flowering 3–6), there are good displays of bluebells in late Spring (4–6) as well as the dainty wood anemone or windflower (4–6). The presence of these last two growing together is usually a sign of old woodland. Marsh-marigold or kingcup brightens damper areas in Spring (3–5), as later do opposite-leaved golden-saxifrage (4–5) and brooklime (5–9).

The reserve is a good place for woodland birds and butterflies, while dragonflies and damselflies are to be seen along the valley pools.

Wood anemone

Management
The ponds were very overgrown but have now been opened up to let in light and air. The retaining dams of one or two of the ponds have fallen into disrepair and we hope to be able to repair them.

Effort concentrates on maintaining and enhancing the variety of habitats in the Spinney, by controlling encroaching rhododendron and willow. The woodland is being improved by thinning and replanting.

SWAINS WOOD

Nearest towns High Wycombe and Henley-on-Thames

An area of 16.6 hectares (41 acres) mainly of woodland but with some chalk grassland. Scheduled as an SSSI, it has been managed by BBONT by agreement with the owner since 1978.

Access
By permit only, obtainable from the BBONT Office in Oxford.

Description
The reserve comprises three main habitat areas. There is an area of mature beech woodland, and also an area of naturally regrown woodland on a site which was clear-felled during World War II. One slope of the dry valley running across the reserve is grassland, with scrub threatening to invade.

Part of this grassland was ploughed during the war but the steeper slopes evaded this fate and carry a marvellous galaxy of wildflowers, those of note being wild orchids (usually at their peak in June) and Chiltern (flowering 8–9) and autumn (8–9) gentians. Both these gentians have purple, trumpet shaped flowers, those of the Chiltern gentian being larger and longer.

There are many butterflies and other invertebrates. Of special note is the purse web spider. As might be guessed from its name, it has an unusual strategy for taking prey. It spins a silk tube in which it lives, the tube partly in the ground and partly lying on the surface camouflaged with scraps of leaf and crumbs of soil. An insect unknowingly walking over or landing on the disguised surface is impaled through the walls of the tube by the spider's giant fangs and paralysed. A hole is torn in the tube and the spider drags the prey in. Before settling down to feed, it repairs the rent in the tube! These spiders are relatively rare but can be quite numerous where they do occur.

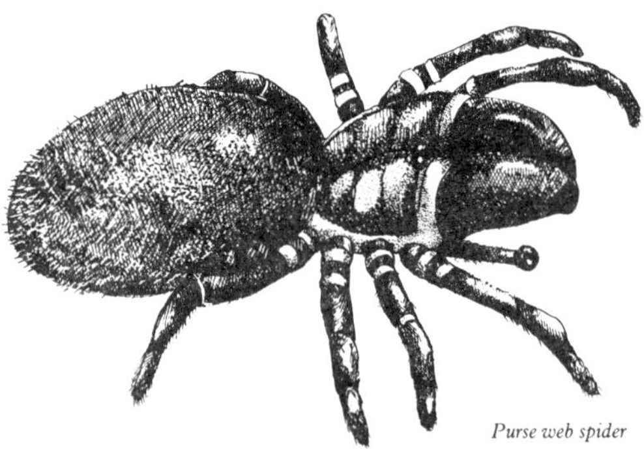

Purse web spider

Management
The owner lets timber rights, and there is some felling and extraction for firewood from time to time. The grassland was fenced by BBONT, who have undertaken much scrub removal and organised sheep grazing to maintain the turf, its wildflowers and wildlife.

OTHER WILDLIFE SITES OF INTEREST

Other wildlife sites in the three counties include:

Ambarrow Hill, Berks SU 825629. ½ mile south of Crowthorne Station. 11 acres of pine clad slopes belonging to the National Trust.

Aston Rowant National Nature Reserve, with Beacon Hill Forest Nature Reserve and Aston Hill (National Trust), Oxon SU 7297. These areas of beechwood and chalk grassland lie astride the M40 between Stokenchurch and Lewknor.

Bagley Wood, Oxon SP 5102. Lying alongside the A34 on the south-west edge of Oxford, this wood is owned by St John's College. Parts of the original mixed woodland remain, but much is planted with conifers.

Bernwood Forest, Bucks and Oxon SP 6009, 6110, 6210 and 6111. Bernwood Forest is the present-day name for the Forestry Commission woodlands lying north-east of Stanton St John. They include fragments of the ancient Royal Forest of Bernwood.

Black Park and Langley Park, Bucks TQ 0083. 4 miles north-east of Slough, Black Park is a country park with various woodlands and a lake, while Langley Park adjoining has some grassland, an arboretum and a farm.

Blenheim Park, Oxon SP 4316. On the edge of Woodstock, this contains fine old oaks, deer, insects and good waterfowl on the lake.

Bradenham Woods, Bucks. National Trust land. Bradenham Wood and Naphill Common lie between Bradenham and Naphill **SU 8397**; Park Wood is ½ mile north of Bradenham SU 9298: a Grade 1 site for plateau and dip-slope beechwood with some oak on Naphill Common.

Bulmershe, Reading, Berks SU 752731. Lying beside Woodlands Avenue is a pond with varied surrounds, managed by the Reading Urban Wildlife Group.

Burghfield Common, Berks – see Wokefield Common

Burnham Beeches and Dorneywood, Bucks SU 9585. ¾ mile west of Farnham Common lie these woodlands famous for their ancient pollard beech trees; they also contain oak and birch. Neighbouring East Burnham Common still has good heathland. Owned by City of London Corporation and the National Trust.

Buscot Park, Oxon SU 239973. Lying between Lechlade and Faringdon, this parkland includes a lake and heronry.

California Country Park, Berks SU 784652. Woodland, heath, bog and lake. Wokingham District Council.

Church Hill, West Wycombe, Bucks SU 828950. ¼ mile west of West Wycombe, a steep hill with chalk grassland, scrub and many old yew trees. National Trust.

Church Wood, Hedgerley, Bucks SU 975875. This is an outlier from Burnham Beeches, owned by the Royal Society for the Protection of Birds.

Clipper Down, Bucks – see Ivinghoe Beacon

Cliveden, Bucks SU 913856. Lying 2½ miles north of Taplow, the well-known house and gardens are supplemented by some fine hanging beechwoods on the cliffs above the Thames. National Trust.

Cock Marsh, Berks SU 886868. 1 mile north of Cookham, 132 acres of marshy meadows plus steep chalk slopes, owned by the National Trust.

Cookham Dean Common, Berks – see Maidenhead Thicket

Cookham Moor, Berks SU 895853. In the middle of Cookham, 9 acres of marshy grassland, owned by the National Trust.

Coombe Hill, Bucks SP 849066. Lying 1½ miles north of Wendover, 106 acres of chalk grassland and scrub with some heathland on the clay-with-flints soil at the top – the summit at 852 ft is the highest point in the Chilterns. Owned by the National Trust.

Devil's Dip, Reading, Berks SU 690724. An old gravel pit managed by the Reading Urban Wildlife Group.

Dinton Pastures Country Park, Berks SU 785718. Flooded gravel pits. Wokingham District Council.

Dorneywood, Bucks – see Burnham Beeches

East Burnham Common, Bucks – see Burnham Beeches

Edgbarrow Woods, Berks SU 837632. Heathland, pine woodland and bog, adjoining BBONT's OWLSMOOR BOG and HEATH reserve. Bracknell Forest Borough Council.

Englemere Pond Nature Trail, Berks. SU 905685. Woodland, lake and heathland. Bracknell Forest Borough Council.

Finchampstead Ridges, Berks SU 808634 and including Heath Pool and Simons Wood. ¾ mile west of Crowthorne Station, 107 acres mainly of conifer woodland and heathland. National Trust.

Great Meadow Pond, Berks – see Windsor Great Park

Heath Pool, Berks – see Finchampstead Ridges

Heron Island, Reading, Berks SU 702719. Once a thriving heronry before being damaged in the October 87 storm. It is being replanted and managed by the Reading Urban Wildlife Group.

High Standing Hill, Berks – see Windsor Forest

Hodgemoor Wood, Bucks SU 9693. 1 mile west of Chalfont St Giles, public footpaths run through extensive woodlands. Forestry Commission.

Hog and Hollowhill Woods, Bucks SU 823860. Near Medmenham, beech woodland – a Bucks CC countryside area with public access.

Hogback Wood, Bucks SU 927912. 1 mile west of Beaconsfield Station, 23 acres of deciduous woodland owned by the National Trust.

Ivinghoe Beacon, Bucks SP 961169. 1½ miles north-east of Ivinghoe; together with Steps Hill and Clipper Down, a fine stretch of chalk grassland owned by the National Trust.

Langley Park, Bucks – see Black Park

Lardon Chase and Lough Down, Berks SU 588809. ¼ mile north-west of Streatley, 66 acres of downland owned by the National Trust.

Lily Hill Park, Berks SU 887694. Ornamental parkland and grassland. Bracknell Forest Borough Council.

Littleworth Common. Bucks SU 935863. 1 mile north-west of Burnham Beeches, heathland, now overgrown, with ponds.

Lodge Hill, Bucks SP 794001. 1½ miles south of Saunderton, astride the Ridgeway Path, an isolated hill with good chalk turf, wildflowers and butterflies.

Lough Down, Berks – see Lardon Chase

Magdalen Meadow, Magdalen College in Oxford SP 524063. A fine fritillary meadow, with a herd of fallow deer in the adjoining park.

Maidenhead Thicket, Pinkney's Green and Cookham Dean Common, Berks SU 855810, 860825 and 863843. West and north-west of Maidenhead, the National Trust here holds 613 acres of woodland, scrub and grasslands.

Moorend Common, Bucks SU 803905. 1 mile south-west of Lane End, a Chiltern common with good oakwood, scrub and grassland.

Naphill Common, Berks – see Bradenham Woods

North Leigh Common, Oxon SP 402140. 1 mile north-east of North Leigh. 41 acres of heathland and scrub, bog and interesting flowers including western gorse.

Northmoor Hill Woods picnic site, Bucks SU 035891. Lying near Denham, this is an area of mixed woodland with alder carr.

Obelisk Pond, Berks – see Windsor Great Park

Park Wood, Bucks – see Bradenham Woods

Pangbourne Meadow, Bucks SU 640768. These 7 acres of Thames-side meadow lie just below Pangbourne Bridge. National Trust.

Pixey Mead, Oxon – see Yarnton Mead

Port Meadow, Oxon SP 4908. This fine Oxford grazing meadow lies 1½ miles north of Carfax.

Prestwood picnic site, Bucks SP 862020. A chalk downland setting for this Bucks CC site.

Rammamere Heath, Bucks – see Stockgrove Park

Rushey Way Pond, Reading, Berks SU 752706. A large pond saved from development by the Reading Urban Wildlife Group.

Shepherds Meadow, Berks SU 850604. Wet meadows next to the River Blackwater. Bracknell Forest Borough Council.

Shotover Country Park, Oxon SU 5606. An Oxford City Council country park which includes a brackeny hillside with some oakwood.

Simons Wood, Berks – see Finchampstead Ridges

Snelsmore Common, Berks SU 460710. 1½ miles north of Newbury, this Newbury DC park contains a fine stretch of heathland and bog.

Somerton Meads, Oxon SP 492302. ¾ mile north-west of Somerton, this is an important haunt for wildfowl.

Stockgrove Country Park with Rammamere Heath, Bucks SP 928300. A mixed woodland and lakes on acid sandy soil. Bucks CC.

Stoke Common, Bucks SU 9885. 1 mile north-east of Stoke Poges, a stretch of overgrown heathland.

Stonesfield Common, Oxon SP 392165. ½ mile south of Stonesfield, some limestone grassland with many wildflowers.

Stonor Park, Oxon SU 740890. ¼ mile north-east of Stonor, an ancient deer park, notable also for its chalkland wildflowers. Owned by the Stonor family.

Stowe Park, Bucks SP 6636. 3 miles north-west of Buckingham, the school grounds include a 25 acre woodland nature reserve.

Sutherland Grange, Berks SU 940772. Rough, unimproved neutral meadow. Royal Borough of Windsor and Maidenhead.

Sutton Courtenay Field Centre, Oxon SU 500920. A 15 acre educational reserve 1½ miles south of Sutton Courtenay and in the grounds of Didcot Power Station.

Swyncombe Downs, Oxon SU 673912. 2½ miles east of Ewelme, this well-known viewpoint over the Thames valley has interesting wildflowers and some juniper.

Taynton Quarry, Oxon SP 239150. 1½ miles north of Taynton, old limestone quarries with a variety of wildflowers and butterflies.

Thatcham Reedbeds, Berks, SU 500668. An area of reedbed, 2 miles west of Thatcham, important for birds and insects. Part is designated a Local Nature Reserve.

Virginia Water, Berks – see Windsor Great Park

Watlington Hill and Park, Oxon SU 702935. 1 mile south-east of Watlington, the National Trust owns 150 acres of beechwoods surrounding Watlington Park and 96 acres of chalk grassland and scrub with many old yew trees.

Wendlebury Meads, Oxon SP 5617. National Nature Reserve 1½ miles north-west of Charlton-on-Otmoor. Lowland meadows with an exceptionally rich flora.

White Horse Hill, Oxon SU 300867. This historic site lies 2 miles south of Uffington. It comprises 235 acres of National Trust land, chalk downland and farmland with outstanding wildflowers.

Widbrook Common, Berks SU 897840. 1 mile south of Cookham, 65 acres of grazed marsh and grassland, owned by the National Trust.

Willen Lake, Bucks SP 8740. This first-class winter waterfowl resort is ¼ mile south of Willen in Milton Keynes.

Windsor Forest, Berks SU 9373. 2½ miles south of Windsor, this fragment of ancient oak forest is especially rich in beetles and other insects. It includes 45 acres of High Standing Wood as a Forest Nature Reserve. Management is by the Crown Estates.

Windsor Great Park, Berks SU 97. Situated 1 mile south of Windsor this includes an extensive stretch of acid grassland, bracken and deciduous woodland, with some fine old oaks and three good waterfowl sites: Great Meadow Pond (965710), Obelisk Pond (977703) and Virginia Water (970690).

Wittenham Clumps, Oxon SU 570925. This well-known viewpoint lies 1 mile south-east of Little Wittenham, where the Nature Reserve is owned and managed by the Northmoor Trust.

Wokefield (or Burghfield) Common, Berks SU 635662. 1 mile north of Mortimer, this is a fine stretch of heathland with alder gullies and good fungi .

Wychwood Forest and Ponds, Oxon SP 3316. 1 mile north-east of Leafield, this remnant of once extensive ancient forest has still some fine patches of oakwood and limestone grassland.

Wytham Estate, Oxon SP 4608. Situated 1 mile west of Wytham, this woodland is used for research by Oxford University biological departments. Access is restricted.

Yarnton Mead, Oxon SP 4711. Lying south-west of Yarnton village, Pixey and Yarnton Meads are important stretches of old flower-rich meadowland managed in the same way for centuries – see the entry for OXEY MEAD (a BBONT reserve).

GIFTS OF RESERVES AND GRANT AID FOR RESERVE PURCHASE

We gratefully acknowledge the gifts of reserves from:

Mr & Mrs E. Hambly (Long Grove Wood)
Mr & Mrs J. A. Lamb (Lamb's Pool)
Mr & Mrs D. Porter (Blenheim Farm)

BBONT thanks the following for their generosity in helping us to purchase reserves.

Miss Diana Alderson
R. Amey, Esq.
Nora, Lady Barlow
Berkshire County Council
Bracknell District Council
Buckinghamshire County Council
Carnegie Trust
Cherwell District Council
Chiltern District Council
Countryside Commission
Crowthorne Parish Council
Ernest Cook Trust
Four Winds Trust
Friends of Vale of Aylesbury
John Paul Getty II KBE
Ms Molly Hyde
Inkpen Parish Council
Milton Keynes Development Corporation
National Heritage Memorial Fund
Nature Conservancy Council
Newbury District Council
Oxfordshire County Council
Pilgrim Trust
Mrs Joanna Robertson
Rothschild Executor & Trustee Company Ltd
Royal Society for Nature Conservation
St Helens Without Parish Council
Sandhurst Town Council
South Oxfordshire District Council
Trustees of Miss W. E. Lawrence
Vale of White Horse District Council
Wilfred & Constance Cave Foundation
World Wide Fund for Nature
And many individual donors

£2 OFF BBONT MEMBERSHIP

If after reading this book you would like to become a member of BBONT simply fill in the coupon below and save £2 on membership.

Not only will you be helping wildlife – you will be helping yourself too:

- ★ **Free magazine 'Natural World' 3 times a year.**
- ★ **Free local wildlife magazine and diary of events 3 times a year.**
- ★ **Free car sticker.**
- ★ **All reserves in this handbook to visit.**
- ★ **Hundreds of organised walks, talks and excursions every year.**
- ★ **Special events and walks for children.**

HELP BBONT HELP WILDLIFE
Fill in the coupon today and help yourself to £2 off membership.

Send to: The Membership Secretary, BBONT, 3 Church Cowley Road, Rose Hill, Oxford, OX4 3JR.
I have purchased a copy of 'Where to Go for Wildlife' and would like to become a member of BBONT
Please tick box ☐

Concessionary

☐ Ordinary £12.50 less £2 ☐ 23 and under £7 less £2
☐ Family £18.00 less £2 ☐ 60 and over £7 less £2

I enclose cheque/PO for £ .....
made payable to BBONT
or debit my VISA/ACCESS card no

☐☐☐☐☐☐☐☐☐☐☐☐☐☐☐☐

Expiry date....................
Signature.....................

Name.........................
Address.......................
....
....
Postcode ....

£2 OFF BBONT *MEMBERSHIP*